# PRACTICAL ELECTRICAL INSTALLATION, REPAIR & REWIRING

**Other TAB books by the author:**

No. 903 *Do-It-Yourselfer's Guide to Modern Energy-Efficient Heating & Cooling Systems*
No. 909 *How To Build Metal/Treasure Locators*
No. 923 *How To Build Your Own Boat From Scratch*
No. 996 *Treasure Hunter's Handbook*
No.1020 *Automotive Air Conditioning Handbook—Installation, Maintenance and Repair*
No. 1126 *55 Easy Outdoor Projects For Do-It-Yourselfers*

No. 925
$12.95

# PRACTICAL ELECTRICAL INSTALLATION, REPAIR & REWIRING

## BY JOHN E. TRAISTER

TAB BOOKS

BLUE RIDGE SUMMIT, PA. 17214

FIRST EDITION

FIRST PRINTING—MAY 1979

Copyright © 1979 by TAB BOOKS

Printed in the United States of America

**Library of Congress Cataloging in Publication Data**

Traister, John E.
    Practical electrical installation, repair and rewiring.

    Includes index.
    1. Electric wiring, Interior—Amateurs' manuals. 2. Household appliances, Electric—Maintenance and repair—Amateurs' manuals. I. Title.
TK9901.T66    644'.3    79-14326
ISBN 0-8306-9776-4
ISBN 0-8306-9774-8 pbk.

# Foreword

You easily will learn how to do professional quality electrical work with this comprehensive answerbook to all your electrical questions. The emphasis is on practicality and simplicity. There's no complicated math to work out. Handy look-it-up formulas and data tables are provided in the back of the book. You'll be able to make electrical repairs or add new installations while you upgrade your electrical skills.

The first four chapters start you out with the basics you need to get going: a quick rundown on tools; a brief course in electricity; modern wiring techniques; and a complete materials rundown. The remaining 20 chapters focus on specific jobs and problems and, most importantly, on troubleshooting.

You get the straight facts on how to get a job done, together with the problem-solving solutions you need to do it. Maintenance is easy once the guesswork is removed and you have all the information you need on testing for and repairing any type of electrical fault.

Easy installation methods and procedures are fully described. Simple and clearly stated step-by-step instructions guide you. And safety procedures are outlined to completely take the risk out of working with electricity so you can get the job done efficiently, inexpensively, and safely.

# Contents

**1** Tools for Home Electrical Work .................................................11
Screwdrivers—Side-Cutting Pliers—Long-Nose Pliers—
Diagonal-Cutting Pliers—Adjustable Wrench—Claw
Hammer—Wood Chisels—Hacksaw Frame & Blades—Step
Ladder—Electric Drill—Voltage Tester—Soldering Iron—
Compass Saw—Folding Rule—Gripping Pliers—Work Safety
Considerations

**2** Understanding Electricity: How It Works...............................23
An Electrical System—A Typical System

**3** Wire Joints ......................................................................30
Stripping & Cleaning Wires—Types of Wire Splices—Wire
Connections—Soldering—Taping Electrical Wire Joints

**4** Electrical Materials for Home & Farm ..................................48
Cable Systems—Raceway Systems—Use of Different Wiring
Systems—Using EMT—Outlet Boxes

**5** Troubleshooting the Home Electrical System .........................61
Types of Electrical Faults—Test Procedures

**6** Adding Duplex Receptacles ................................................71
Plan Ahead—Concealed Wiring in Accessible Spaces—
Calculating Receptacles Per Circuit

**7** Installing Light Switches ...................................................88
Installation—Installing Surface Metal Molding—Installing a Wall
Switch in Plastered Wall—Other Switch Applications

**8** Why Walk in the Dark? ............................................... 105
Three-Way Switches—Four-Way Switches—Three-Way Switching System Installation—Low-Voltage Remote-Controlled Switching—Application of Low-Voltage Switching

**9** Ready to Tackle 240-Volt Circuits? ........................... 119
Layout of 240-Volt Circuits—Practical Applications

**10** Update Your Own Electric Service ............................ 132
Correct Size of Electric Service—Installation of Service Equipment—Pole Metering—Partial Updating of Electrical Service

**11** Grounding—A Life Saver .......................................... 152
Equipment Grounding—Portable Appliances & Hand Tools—Summary of Equipment Grounding

**12** Lighting the Home's Interior ................................... 164
Selecting Lighting Fixtures—Installation of Lighting Fixtures

**13** Outdoor Lighting .................................................... 175
Installing Outdoor Farm Lighting—Installing Outdoor Residential Lighting

**14** What About Motors? ............................................... 188
Capacitor-Start Motor—Split-Phase Motor—Capacitor-Start Capacitor-Run Motor—Repulsion-Start Induction-Run Motor—Universal Motor—Shaded-Pole Motor—Electric Motor Maintenance—Typical Applications of Electric Motors—Motor Repairs

**15** Why Overcurrent Protection? ................................. 199
Plug Fuses—Cartridge Fuses—Circuit Breakers—Selecting Proper Overcurrent Protection

**16** Lightning Protection for the Home & Farm .............. 207
Lightning Rods—Lightning Arresters

**17** Television Systems in the Home .............................. 221
Antenna Head—Preamplifier—Rotor—Masts—Lead-In Cable—Couplers—Antenna Signal Splitter—TV Outlets—Master Antenna TV System

**18** Music All Through the House .................................. 233
Careful Planning—Installation

**19** Notice Your Guest .................................................. 244
Location of Chimes—Types of Chimes—Installation of a Two-Note Chime

**20** Security Systems .................................................... 252
Smoke Detection Alarms—Surface-Mounted Fire/Security Systems

**21** Built-In Central Cleaning Systems .................................... 262
Planning the Tubing System—Installing Tubing & Fittings—Final
Checkout

**22** Installing Add-On Air Conditioning .................................. 289
Mounting the Condensing Unit—Installing the Cooling Coil—
Installing the Refrigerant Tubing—Installing the Electrical Wiring

**23** Electronic Garage Door Controls .................................... 299
Why Have One?—Installation—Optional Accessories

**24** Major Appliance Considerations .................................... 306
Compactor—Disposal—Refrigerator & Freezer—Clothes
Washer—Clothes Dryer—Room Heater—Water Pump—Waste
Aerator—Room Air Conditioner—Dishwasher—Sump Pump—
Barbecue & Smoker—Vent Fan—Food Center—Wall Toaster—
Firelogs—Humidifier—Evaporative Cooler—Microwave Oven

Glossary .................................................................... 319

**Appendices**

I—Manufacturers of Residential Electrical Products .......... 330

II—Wiring Tables .......................................................... 336

III—Electrical Formulas ................................................ 343

IV—Lamp Data ............................................................ 349

Index ........................................................................ 399

# Chapter 1
# Tools for Home Electrical Work

Some homeowners already have a home workshop with some equipment and tools suitable for work on electrical systems. Others will have to start from scratch. But whatever the situation, the homeowner who wishes to work on his own electrical system need not put a strain on his pocketbook. He can purchase the few tools most necessary and add to them as the need arises.

In general, tools of high quality are the cheapest in the long run. If the homeowner selects tools of good quality, uses them properly and takes good care of them most will usually last a lifetime. A list of the more common and necessary ones for home maintenance are as follows:

- Several screwdrivers of various sizes
- Side-cutting pliers
- Eight-inch diagonal cutting pliers
- Long-nosed pliers
- Adjustable wrench
- Claw hammer
- Wood chisels, one narrow and one wide
- Hacksaw frame and blades
- Step ladder
- Quarter-inch electric drill
- Assortment of drill bits

- Voltage tester
- Soldering iron
- Compass saw
- Six-foot folding rule
- Gripping pliers commonly known as Channel Locks

In addition to this list, one should obtain an ohmmeter tester, which may be a combination volt-ohm-milliammeter. Since the ohmmeter measures resistance in an electrical circuit it can be used to locate many electrical defects. An inexpensive multimeter will cost approximately $20 to $30. This may have to wait until a definite need arises before the purchase would be warranted.

## SCREWDRIVERS

The ordinary screwdriver with a steel shank and wood or plastic handle is fine for home electrical repair work. They are classified by size, according to the combined length of the shank and blade. In shank sizes they run 2½, 3, 4, 5, 6, 8, 10 and 12 inches. The diameter or thickness of the shank and the width and thickness of the blade tip which fits into the screw slot are proportional to the length of the shank. A 6-inch and a 10-inch screwdriver should take care of the majority of home electrical repairs. However, always select the size of screwdriver so that the thickness of the blade makes a good fit in the screw slot even if you have to make a quick trip to the local hardware store to purchase a new one. A good fit not only prevents the screw slot from becoming burred and the blade from being damaged, but also reduces the force required to keep the screwdriver in the slot.

Phillips-type screw heads have become very popular in recent years especially on housing and different forms of trim on electrical apparatus. The heads of these screws have two slots which cross at the center; this prevents the screwdriver from sliding sideways out of the slot and marring the finish of the trim, etc. Three shank lengths of Phillips screwdrivers—4, 6 and 8 inches, will handle all Phillips-head screws commonly encountered around the home. A typical Phillips screwdriver and screw are shown in Fig. 1-1. Common Phillips sizes range between 0 and 3 with sizes 1 and 2 being the most common.

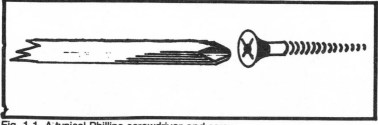
Fig. 1-1. A typical Phillips screwdriver and screw.

## SIDE-CUTTING PLIERS

Since most of the homeowner's electrical work will consist of wire cutting and wire connections, he will need a pair of 6- or 9-inch regular side-cutting pliers. These are commonly known as linemen's pliers, and they have a cutting edge on the side. They are used to cut wire and cable and also to make splices, that is, to twist the ends of two wires together to form a connection. Many people also use these pliers to strip insulation from wire.

In purchasing a pair of side-cutting pliers make certain that the handles are insulated to provide protection against electrical shock. Also hold the pliers up to the light and look through the cutting edge to make certain that the cutting edges meet throughout their entire length. Unless the cutting edges meet in this manner the pliers are no good for cutting wire.

Pliers, like all other tools, should be kept clean. An occasional drop of oil on the joint pin will keep the pliers working freely and easy to use.

## LONG-NOSE PLIERS

Another type of pliers used in electrical work is the long-nose pliers. A pair of 8-inch long-nose pliers should be perfect for electrical work. Again, make certain that the handles are insulated.

Long-nose pliers are useful for recovering washers, nuts, etc. that get into a place where it is hard to reach. They are also ideal for making wire eyes (see Fig. 1-2) for connecting to a screw connector on a receptacle or light switch. However, long-nose pliers should never be used to twist leads of wire together since the jaws of the pliers will eventu-

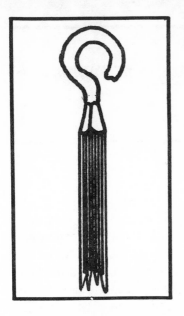

Fig. 1-2. A wire eye for connecting to a screw terminal.

ally become out of alignment and the points will not meet. The twisting of leads together is a job for the linemen's pliers.

## DIAGONAL-CUTTING PLIERS

Diagonal-cutting pliers or dikes are good for trimming wires close to the point of connection so that loose ends cannot come in contact with other wires or grounded circuits.

## ADJUSTABLE WRENCH

Adjustable wrenches are very similar to the common open-end wrench except for having one adjustable jaw. This type of wrench is often called a Crescent wrench. An 8-inch size should be just about right for the homeowner.

Adjustable wrenches are not intended for hard service like the open-end wrenches. Whenever it becomes necessary to exert any amount of force on an adjustable wrench to tighten or break loose a nut always place the wrench on the nut so that the pulling force is applied to the stationary jaw side of the handle. After placing the wrench on the nut tighten the adjusting knurl so that the wrench fits the nut snugly. Figure 1-3 shows the right and wrong way to use an adjustable wrench.

## CLAW HAMMER

There are many types of hammers, but for home electrical work, a straight-claw hammer, sometimes called electrician's hammer, will be the most useful.

The hammer handle should always be tight in the head. Never work with a hammer having a loose head as the head

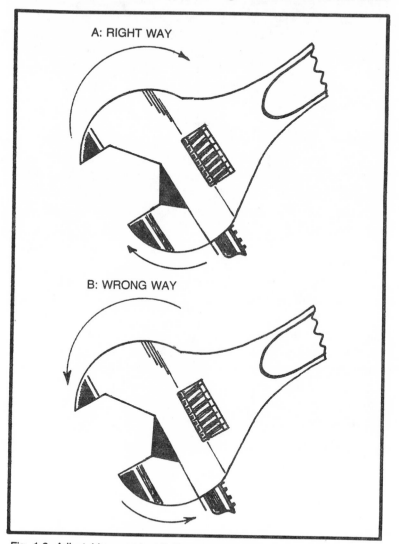

Fig. 1-3. Adjustable wrench. (A) Correct way of using an adjustable end or crescent wrench. Notice that the pressure is applied to the nut by the wrench and not the adjustable jaw. (B) Incorrect method of using the adjustable end wrench. This time pressure is applied to the adjustable jaw.

may fly off and either injure someone or damage surrounding items.

## WOOD CHISELS

Wood chisels usually are used in electrical work in notch beams, joists, studs, etc. for outlet boxes, cable, and conduit. A ¾-inch blade will be used for wire or cable notches while a 2-inch blade will find use in notching wood for outlet boxes. The cutting edge of a chisel must of course be sharp in order to cut, and the sharpness is best maintained by buying wood chisels of high quality and using them only for jobs which they are intended.

## HACKSAW FRAME & BLADES

The homeowner will find many uses for a good hacksaw; cutting armored cable, electrical conduit, etc. A cheap hacksaw frame is next to worthless so only the best should be purchased. Only the better quality blades will hold up for all kinds of electrical work.

## STEP LADDER

Many homeowners will invariably use a rickety old chair, box, or some other unsafe means to change a lamp or to do other overhead work that cannot be reached from floor level. To continue this practice is certain to make it necessary to collect on the person's hospitalization insurance. A sturdy 6-foot step ladder is not expensive and can save a broken leg or a more serious accident. Buy only wood or fiberglass ladders for electrical work and never paint any ladder other than with clear varnish.

## ELECTRIC DRILL

A quarter-inch electric drill is desirable for boring holes in studs and joists to accommodate electrical cable. Some electric drill suppliers furnish complete kits containing all sorts of attachments fitting the drill. Some also include a small stand which converts the small drill into a drill press, which is useful for drilling holes in outlet boxes or other components. Drill bits should include three high-speed wood boring drills—

½-inch, ¾-inch and 1¼-inch—and a set of high-speed metal boring drills from ⅛-inch to ½-inch.

A variable-speed drill will be the most useful especially if a set of masonry drill bits are also included.

## VOLTAGE TESTER

The first rule when working with electricity is to disconnect the circuit on which you plan to work. However, this is not always foolproof in preventing electric shock because the wrong fuse or circuit breaker may be disconnected by mistake. Therefore, a voltage tester is absolutely necessary to assure that no voltage is present in the circuit on which you are working.

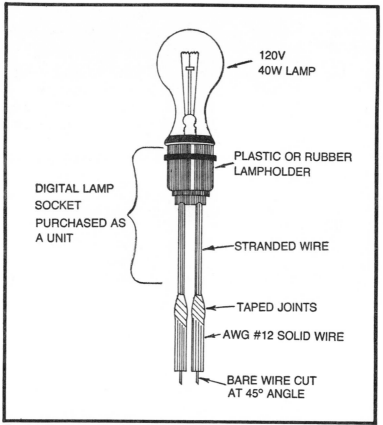

Fig. 1-4. Simple test lamp made from a pig tail lamp socket, a few feet of wire and two test probes.

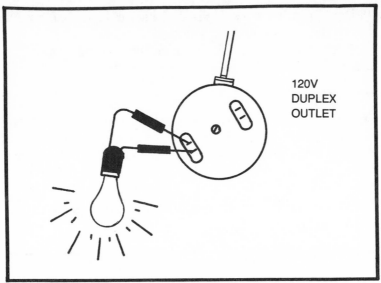
Fig. 1-5. Method of testing a duplex receptacle for voltage.

Figure 1-4 shows a simple test lamp that you can make. It consists of a 40-watt lamp, a pigtail lamp socket, and a few feet of wire. With this test lamp you can test for the presence of voltage, thus locate an open faulty circuit or else insure that the circuit you wish to work on is disconnected.

To make this test lamp you will need two lengths of AWG #12 insulated wire. Remove a half-inch of insulation from one end of each wire leaving the bare wire exposed. Then remove ¾-inch of insulation from the other end of each wire, and bend an eye or hook in this end.

Remove ¾-inch of insulation from both lead wires from the pigtail lamp socket. Wrap one of the bare leads around the hook on one length of the AWG #12 wire, then the other. Squeeze the hook on each lead closed then solder and tape the connection.

Once the lamp is screwed into the socket the test lamp is ready for operation. The remaining half-inch of bare wire on the two ends will be used for testing live electrical components and should never be touched; always grip the test leads well back on the insulated part.

To see how effective this test lamp can be insert one lead into one slot of a 120-volt AC receptacle, then the other lead into the opposite slot of the receptacle (see Fig. 1-5). If voltage is present, the 40-watt lamp will light.

If you plan to work only on 120-volt circuits then the test lamp just described will be sufficient. However, it cannot be used in its present form to test 240-volt circuits as the lamp will burn out immediately if subjected to voltage much higher than 120 volts. But this lamp may be modified as shown in Fig. 1-6 for use on 120 or 240 volts. The only difference in the two is that two pigtail lamp sockets are connected in series using two 40-watt lamps. When the test lamp is used on 120-volt circuits the two lamps will burn very dim; when 240-volt circuits are tested, the lamps will burn to full brilliance.

As mentioned earlier, you may want to purchase a volt-ohm-milliammeter or multimeter, especially if you intend to do much troubleshooting of your own electrical system. Such a tester is shown in Fig. 1-7. These multimeters have several scales across the meter face that correspond to voltage, current and resistance. Many electrical tests may be performed with this type of instrument that will be described in later sections of this book.

## SOLDERING IRON

While solderless connectors will be used for the majority of modern electrical repairs it sometimes becomes necessary

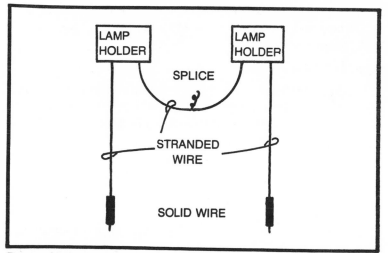

Fig. 1-6. Modified version for use on either 120- or 240-volt circuits.

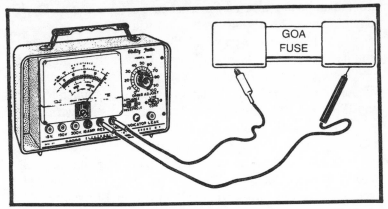

Fig. 1-7. Voltmeter-ohmmeter that may be used to perform many of the tests described in this book.

to solder various electrical connections. The soldering irons have electric heaters built into them. When the iron is plugged into a receptacle, current flows through the heating element. The heat is developed and is conducted to a copper tip. The tip is held against the joint to be soldered.

The soldering gun is a variation of the soldering iron. Both are used for the same type of work, but instead of a heating element and a separate tip in soldering irons, the soldering gun uses a loop of copper or copper alloy which acts as the heating element and the tip. A transformer in the gun's case supplies high current at low voltage which permits the gun to heat and cool quickly. This is a definite advantage over the soldering iron.

## COMPASS SAW

When new outlets need to be added around the home, the compass saw usually is the tool used to cut out the opening. It can also be used to notch wood studs and joists. The homeowner will find many uses for the compass saw. Since the cost of the best is very low get the highest quality possible with a couple of equally find blades.

## FOLDING RULE

This item needs no introduction as the need of measuring is obvious. Again, always buy a good quality rule and take good care of it. You will find yourself needing it quite frequently on nearly every electrical job you perform around the house.

## GRIPPING PLIERS

These pliers can be used as a wrench if the work load is not too heavy and can speed up many operations. One use is to tighten pipe threads when using conduit (pipe) to enclose the electrical conductors. They also come in handy when any type of gripping is needed or for tightening nuts, etc.

Many homeowners have a workshop in their homes, such as in a vacant room, basement, attic or outbuilding. This would be a fine location to store your tools, but unlike a project such as building a bookcase or coffee table in a home workshop, electrical repairs require that the tools be moved to the location of the repair. Therefore, a toolbox or pouch becomes necessary to transport your tools from room to room where the repairs are to be made.

A leather pouch is recommended for small hand tools, such as screwdrivers, pliers, etc., while a suitable toolbox is recommended for the larger and heavier tools, such as a hammer, drill motor, and compass saw. This allows the proper tool to be located with little difficulty and also protects the tools against damage.

Looking back over this chapter the tools recommended may seem excessive to some. But these are the minimum with which the homeowner can perform the common phases of electrical repairs. In almost any project the reader attempts he will find, in a very short period of time, all the tools mentioned will be used.

The tools mentioned will also come in handy for other home repairs such as fixing a leaking faucet, repacking a plumbing valve, etc. So the tools will not be used solely for electrical work.

## WORK SAFETY CONSIDERATIONS

There are common safety rules which anyone who uses electricity should follow, but there are additional rules which every homeowner must consider every time he works on an electrical repair job. Here are a few basic principles he must keep in mind.

- Overloading a circuit is the major cause of trouble in electrical systems in the home. Overloading is dangerous because it not only causes electrical failures, but also can cause fires if improperly fused.
- Never work on any electrical circuit when it's hot; always disconnect the power, then check with a voltage tester to double-check that you disconnected the correct circuit.
- Avoid working in cramped surroundings where, if you did get an electrical shock, you would not be able to get away from shock source immediately.
- Never grab any electrical wire even if you are sure that no voltage is present without first testing it lightly with you finger, keeping your other hand in a pocket. This may seem ridiculous to you, but this gives us a little more insurance to prevent dangerous electrical shock.
- Never fuse an electrical circuit greater than the wire current-carrying capacity.
- Electrical connections should be secure so they will not work loose with vibration or normal use.
- Take your time. Hurry reduces caution and invites accidents.
- When working on concrete floors, outside or in similar areas where your body becomes grounded when it contacts the surface on which you are standing, use a wooden board to stand on or else wear rubber-soled shoes.

# Chapter 2
# Understanding
# Electricity: How It Works

Electricity is the major source of energy for operating many appliances and equipment in the home and around the farm. Yet, in most cases, most people take this form of energy for granted. They know that all they have to do is turn a switch on an electric range to cook, flip a wall switch for light or turn up a thermostat for heat. There are times, however, when the flipping of a switch may produce only a slight click with no other reaction. This is the time when it pays the homeowner to have a basic working knowledge of electricity so that the problem can be corrected.

Rather than get into a comprehensive study of the electron theory or nature of electricity, we will be more concerned with what electricity will do for us, how it can be controlled, how to install electrical equipment and what to do when things go wrong.

First, in order for electricity to be useful, it must be in motion, that is, dynamic electricity. Static electricity (standing still), except in a few cases, is no more useful to us than water standing still in a garden hose. The faucet may be turned on like an electric generator operating to produce electricity, but if the nozzle is closed (like no load connected to the generator) the water is not in motion and is useless until the nozzle is opened to let the water out. Likewise, the

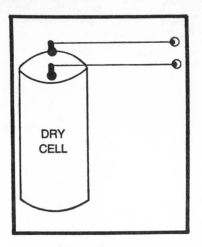

Fig. 2-1. An electrical circuit connected to a potential voltage source with no load connected.

electricity is not in motion until some electrical load is connected to the circuit to produce a flow of current.

For example the circuit in Fig. 2-1 shows a battery connected to a two-wire circuit. An electrical potential is present, but since no load is connected to the circuit no current flows through the wires. However, if an electric lamp is connected to the circuit as shown in Fig. 2-2, an electric current will flow to produce useful electricity and light the lamp.

When electricity is in motion it provides the most effective means known to date for carrying energy from one place to another, and of changing one form of energy to another form of energy.

To illustrate burning coal is a form of heat energy that may be used to change water into steam to drive engines that produce energy of motion. These steam engines are used to drive electric generators which convert this mechanical energy into electrical energy. The electric energy is transported for great distances to factories, houses and farms where the electric energy is converted into many other forms of energy like light, heat, mechanical motion, chemical, sound and radiation.

By now you should see that electricity is the universal means for changing one form of energy into other kinds. It is the only means by which we may transmit power in large quantities from where it is cheaply or conveniently produced to somewhere else many miles away where the power may be used to an advantage.

## AN ELECTRICAL SYSTEM

In general, an electrical system may be classified into one of several groups and may be defined as follows:

■ **Source**—the source of energy to drive the devices (generators) to convert other forms of energy into electric energy may include mechanical, heat, radiation and chemical

■ **Transformer**—devices to increase the voltage at the generators so that the current may be transmitted over long distances with the least amount of loss

■ **Transmission Lines**—this group covers all the conductors that carry the moving electricity from place to place like from the generator or powerhouse to your city. It also includes the insulation or insulators which prevent the escape of electricity from the conductors

■ **Controlling Mechanisms**—this group consists mainly of manual and automatic electric switches of several varieties

■ **Step-Down Transformer**—device to reduce the high voltage of the transmission lines to a lower usable voltage at the location of utilization

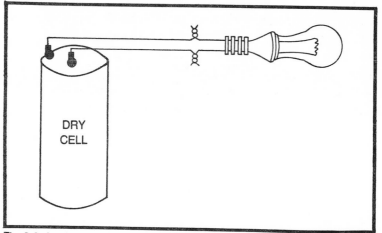

DRY
CELL

Fig. 2-2. An electric lamp connected to the circuit shown in Fig. 2-1 producing a flow of electrical current.

■ **Meters**—instruments to measure the amount of current used at any given location.

■ **Point of Utilization**—at this point, apparatus is provided for changing the moving electricity into some other form of energy which can be used. In this group we find motors (mechanical energy), storage batteries (chemical energy), resistance heaters, lamps, etc.

## A TYPICAL SYSTEM

A typical electrical system is outlined in Fig. 2-3. In this illustration, some form of energy other than electricity is used at the generating station to drive the generator. This energy could be produced by coal, gas, water or atoms to name a few.

Let's assume that the electricity is generated at a pressure of 11,000 volts. Before it reaches the transmission line it is run through a transformer to increase the voltage to 138,000 volts so that the moving electricity can be transmitted over long distances. It finally reaches an area of utilization such as a large industrial installation or a town or city substation. Here the voltage is reduced somewhat; this could be from 2300 to 15,000 volts.

The reduced voltage then continues over local power lines to the various buildings requiring electric service. At these points transformers are again used to reduce the voltage to a usable safer rate. For example, most farm and residential applications will have three-wire electric service of 120/240 volts.

The electric service is run through an electric meter before entering the building. Various circuits are utilized to carry the moving electricity to lights, receptacles and other outlets where the electricity is used or converted to some other form of energy.

Since it is important for any homeowner to know whether his home has an electrical system that is adequate for carrying current to the many appliances and machines or not we will dwell on this last category somewhat.

The drawing in Fig. 2-4 will give you an idea of what constitutes adequate wiring in the average home.

First there should be three lead-in wires from the power company's pole to your home. This indicates that your service

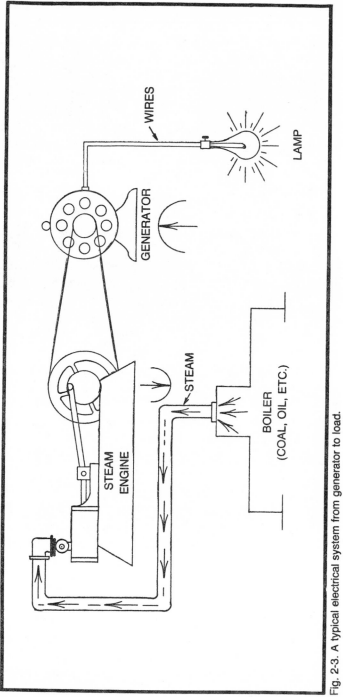

Fig. 2-3. A typical electrical system from generator to load.

entrance is 120/240 volts. These service conductors will travel through the power company's meter to either a main switch or main circuit breaker panelboard. The ampere rating of either should be at least 100 amperes. If it is less, your electrical system is probably overloaded.

Actually if you are planning to update your electrical service a 200-ampere service is none too large. A service of this size should handle all of your present electrical needs as well as any future requirements. Of course, there are exceptions. Some large all-electric homes may require an electrical service of 800 amperes or more.

From the main switch or panelboard there will be several 120-volt appliances. It is recommended that all branch circuits have a minimum wire size of AWG #12, but don't get upset if some of them are AWG #14. The National Electrical Code permits the use of AWG #14 on all residential circuits except those designed to carry small appliance loads as will be described later. Still, to keep voltage drop (reduction in voltage at receptacle due to too small of wire gauge, etc.) to a minimum most designers and contractors will recommend using AWG #12 throughout the home.

You will also find 240-volt circuits feeding such items as your electric range, electric heat, water heaters, clothes dryer, etc. The size wire used for these circuits will vary from AWG #12 AWG to AWG #6 or higher (the smaller the number the larger the wire size).

Regardless of the circuit or type of electrical load connected, all should be protected with some form of over current protection like a fuse or circuit breaker. Proper splices, insulation, switches, and proper grounding are other points to be considered for adequate wiring and all of these will be covered in later chapters. However, before getting into the finer points of your electrical system, the following list gives some signs to look for in the home to determine if your present wiring is inadequate:

- Fuses blow or circuit breakers trip often
- Outlets and light switches seem scarce when you need them
- Lights dim or flicker when heavy appliances are used

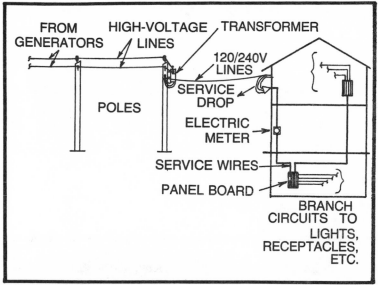

FROM GENERATORS — HIGH-VOLTAGE LINES — TRANSFORMER — 120/240V LINES — SERVICE DROP — POLES — ELECTRIC METER — SERVICE WIRES — PANEL BOARD — BRANCH CIRCUITS TO LIGHTS, RECEPTACLES, ETC.

Fig. 2-4. An example of what constitutes adequate wiring in the home.

■ Television picture shrinks when heavy appliances are used

■ Extension cords are in common usage around the home

■ Appliances are slow starting and slow operating, plus electric motors overheat easily

If you have noticed any of the preceding signs occurring in your home you should have the wiring checked immediately, either by following instructions given in this book or by having your local electrical contractor check your wiring.

# Chapter 3
# Wire Joints

In working with wires for any electrical system or device it is necessary to make numerous connections and splices of various types. Therefore anyone who anticipates working with electricity should have a good knowledge of wire connectors, splicing and soldering. Splices and connections that are properly made will last as long as the wire itself or its insulation, while poorly made connections will always be a source of trouble and will overheat, injure their insulation and cause high resistance in the circuit which could cause injury to life and property.

The requirements for a good electrical connection are it should be mechanically and electrically secure and continue as such as long as the circuit is in operation, that is, it should be electrically sound to carry the required load without overheating, and mechanically strong enough to stand as much strain or pull as the conductor or wire itself.

There are many different types of joints for different purposes, and the selection of the proper type for a given application will depend to a great extent on how and where the wire is used. The types most commonly used in the home around the farm is fully explained in this chapter.

## STRIPPING & CLEANING WIRES

The first step in making any splice or connection is to properly strip and clean the ends of the wire. Stripping means

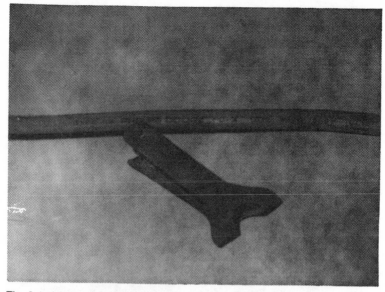

Fig. 3-1. A tool used for stripping sheathing from type NM cable.

removing the insulation from the wire a distance from between 1 to 4 inches depending upon the size of wire and the kind of splice or connection to be made. While experienced electricians can do very well with a pocket knife or a pair of side-cutting pliers in stripping wires of any size the homeowner will find one of the special tools designed for this purpose a great aid in stripping wires, especially those sizes smaller than AWG #10. All such tools are inexpensive and are constructed to prevent cuts or nicks in the wire. Cuts or nicks in conductors reduce the conducting areas as well as making the wire weaker at the point of the nick.

Figure 3-1 shows the use of one type of stripping tool in stripping sheathing from type NM (Romex) cable. The stripper is first inserting on the cable back from the end at the desired distance, usually 8 to 10 inches for inserting in outlet boxes. Pressure is then applied to both sides of the tool with the thumb and forefinger. Then while holding the cable tightly in one hand, the stripper is pulled towards the end of the cable as shown in Fig. 3-2. The cut sheathing is peeled back and cut neatly with either a pocket knife or a pair of cutting pliers (see Fig. 3-3).

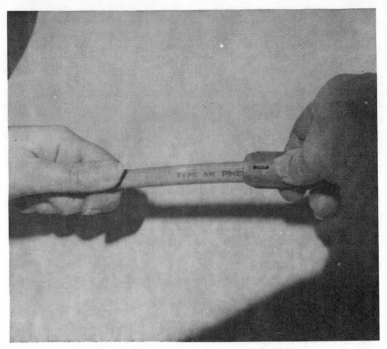

Fig. 3-2. Illustration showing method of use of the tool in Fig. 3-1.

Once the sheathing has been removed from the cable, the individual conductors may be stripped with the tool as shown in Fig. 3-4. This tool has several round openings for various wire sizes; select the proper opening, clamp the tool on the conductor at the desires distance from the end and while holding the conductor tightly in one hand pull the tool towards the end of the wire which will remove the insulation.

The one-shot stripping tool in Fig. 3-5 can be a great time-saving device when many wires need to be stripped. Merely place the end of the wire in the jaws of the tool and clamp down on the handles in a way similar to using a pair of pliers. The insulation is stripped from the wire immediately in one operation.

The stripping tools previously described are usually manufactured for wire sizes AWG #10 and smaller. Although stripping tools are manufactured for the larger size wires their cost makes them impractical for the average homeowner. Therefore, a pocket knife should be used for stripping the

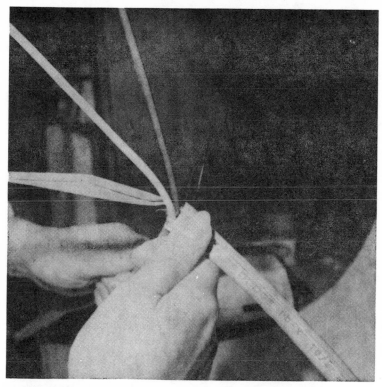

Fig. 3-3. Method of removing sheathing from NM Cable.

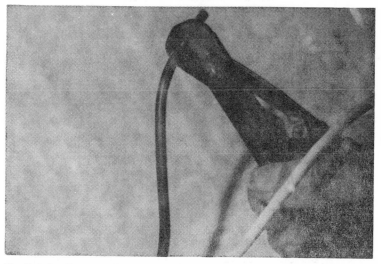

Fig. 3-4. Another type of tool used to remove insulation from individual conductors in the cable.

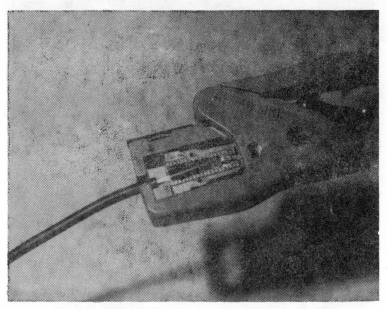

Fig. 3-5. Use of a one-shot tool for removing insulation from conductors.

larger size conductors. The knife, as shown in Fig. 3-6, should be held in a position similar to that used when sharpening a pencil, and the insulation cut through at an angle as shown.

Never cut the insulation straight through as it is very easy to nick the wire when cutting in this manner; it sometimes makes a more difficult splice to properly tape.

After cutting through the insulation and down to the wire let the blade slide along the wire, stripping the insulation to the end. Several passes may be necessary to remove all of the insulation. In doing so, keep the blade side of the knife almost flat against the wire so it does not cut into the metal.

Once the insulation is removed from the conductor the wire should be scraped with the back of the blade to remove all traces of insulation. Do this until the wire is thoroughly clean and bright as this insures a good electrical contact as well as insuring that solder will adhere readily. If the wire is tinned do not scrape deep enough to remove the tinning if the joint is to be soldered. Rather, leave on as much as possible as it makes soldering easier. When cleaning the ends of the wires, emery cloth may be used in place of the back of the knife blade for a neater job.

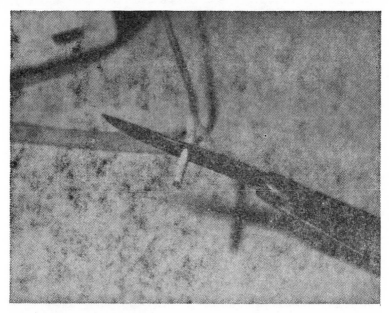

Fig. 3-6. Position of knife when used to remove insulation from conductors.

## TYPES OF WIRE SPLICES

There are dozens of different wire splices for different uses, but for the wiring repairs and additions around the farm or in the home four types should suffice. They are rattail, Western Union, simple tap and fixture-wire splice.

The rattail splice (Fig. 3-7) is the type most often used to splice conductors in outlet boxes. To make this splice strip and clean about 1½ inches on the end of each wire. Then twist them together a few turns with your fingers. Finish the turns with a pair of pliers until the twists are very tightly wrapped

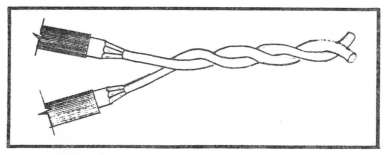

Fig. 3-7. Rattail splice.

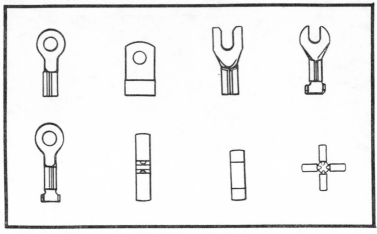

Fig. 3-8. Illustration of several types of wire connectors.

around each other. Be certain that one wire does not remain straight while the other wraps around it as this will enable the wires to slip or shift upon each other. In making this or any other splice the splice itself should be tight and strong before any type of wire connector or solder is applied.

After the rattail splice is completed it may be soldered and taped as will be described later. However, for this type of splice various wire connectors (see Fig. 3-8) are much faster than soldering and are highly recommended for rattail splices in outlet boxes where there is no strain on the wires.

If the wires are stripped to the proper length before the wire nut is applied no further insulation will be required as can be seen in Fig. 3-9. However, in Fig. 3-10 the insulation was stripped too much and bare unprotected wires remain exposed. When this occurs the wire nut should be removed and the wire recut to allow the wire nut to fit properly with no bare wires showing. For added protection, a small piece of electrical tape may be applied to the wire nut around the opening on the bottom, but usually this is not necessary.

When a tap or branch wire is to be connected to a feeder wire, such as overhead wires running from building to building around the farm, a simple tap splice as shown in Fig. 3-11 may be used. For this splice strip about 1 inch of insulation from the main feeder or running wire at the point of the tap, and about 3 inches on the end of the tap or branch wire. The stripped wire

36

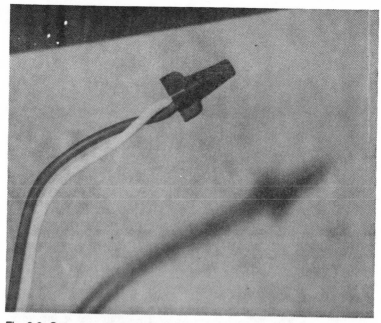

Fig. 3-9. Correct method of securing conductors with a wire nut.

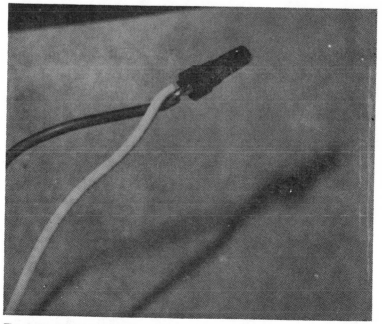

Fig. 3-10. Incorrect way of securing conductors with a wire nut; notice the exposed wires.

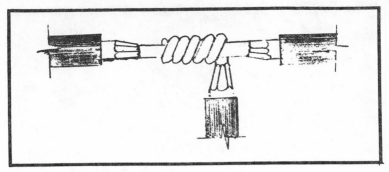

Fig. 3-11. A simple tap splice.

should be cleaned as described earlier before wrapping the 3 inches of tap wire tightly around the main feeder wire. Make approximately from five to eight turns around the feeder as shown in Fig. 3-11. The wires should be tight enough so that they will not slide or slip.

After the splice is made the joint can be soldered and taped, but a solderless lug, such as the split bolt connector shown in Fig. 3-12, may be used instead for a faster and simpler connection. Once the splice is wrapped, insert the split bolt connector over the splice, tighten with wrenches, then tape. If this or any other splice is exposed to outside weather conditions the tape should be of the all-weather type.

For splicing straight runs of wire such as overhead feeder wires around the farm or various appliance cords in the home, the Western Union splice is one of the oldest and most commonly used. It is very strong and can withstand considerable strain and pull. This type of splice can be used on both large and small wires.

To begin the Western Union splice strip and clean approximately 4 inches of insulation on the end of each wire. Then hold the ends together tightly with your hands or pliers as shown in Fig. 3-13A. Continue twisting the wires together a couple of turns as in Fig. 3-13B. Then wrap the end of each wire around the other wire in five or six neat and tightly wrapped turns as shown in Fig. 3-13C. The splice is completed by clipping the ears or ends of the wires, then pinching them down tightly with pliers. The finished splice should appear as shown in Fig. 3-13D.

Fig. 3-12. A tap made with a solderless lug.

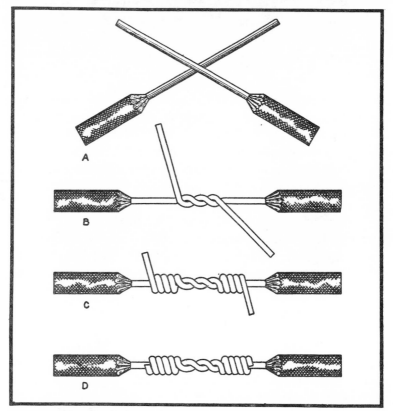

Fig. 3-13. Procedure for making a Western Union splice.

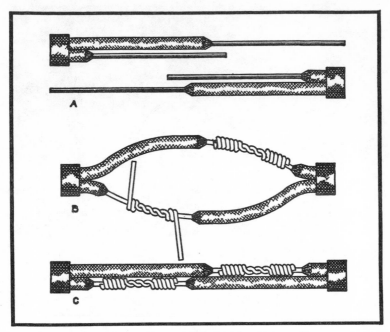

Fig. 3-14. Procedure for making a double Western Union splice.

When making a double Western Union splice such as in an appliance cord, always stagger the joints so that each splice lies next to original insulation as shown in Fig. 3-14. Figure 3-14A shows how the ends of the wires are stripped unevenly, then Fig. 3-14B how the wires are spread to make run for the splices; finally Fig. 3-14C shows how the finished splice appears prior to soldering a taping.

Factory-wired light fixtures normally are furnished with stranded fixture wire smaller than the branch circuit feeding the lighting outlet. At times, a tight splice is very difficult with wires of different sizes, especially if one of the wires is stranded. A fixture splice as shown in Fig. 3-15 will normally provide a good splice after soldering or when a wire nut is installed.

This type of splice is made by wrapping the smaller wire around the larger wire a few turns. Next the larger wire is bent back over the turns in the smaller wire. Then finish wrapping the smaller wire around both sections of the larger wire before soldering or using a solderless wire connector.

## WIRE CONNECTIONS

We have just covered splices, that is, the joining of two or more pieces of wire together. Wire connections are used to connect a wire to an electrical device, such as as receptacle, wall switch, pump control switch, etc.

The homeowner will probably encounter wiring devices with screw terminals more often than any other type. The simple eye connection is the one to use for such terminals. To make the eye in the wire, strip and clean approximately 1 to 1½ inches of insulation from the end of the wire first. With your long-nose pliers, make a slight bend in the wire near the insulation and at an angle of approximately 45 degrees. Continue by bending the wire (above the first bend) in the opposite direction and at different points to form a circle in the wire as shown in Fig. 3-16. The eye may then be placed under the screw terminal and tightened. Always place the eye under the screw head so that the direction of the second bend in the wire is the same as the direction the screw will be tightened. This will cause the eye to close tightly around the screw threads. If the eye is reversed the eye will open and be loose around the threads (see Fig. 3-17).

One type of wire connectors is shown in Fig. 3-18. Some of these are used for connecting wires to screw terminals. To install on a wire strip and clean a length of insulation from the wire exactly the length of the slot on the connector. Insert the bare end of the wire in the open slot, then crimp the slot down tightly against the wire with a pair of pliers or crimping tool. With the wire now secure in the connector the connectors eye

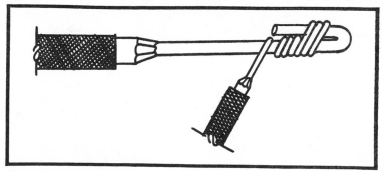

Fig. 3-15. A fixture splice.

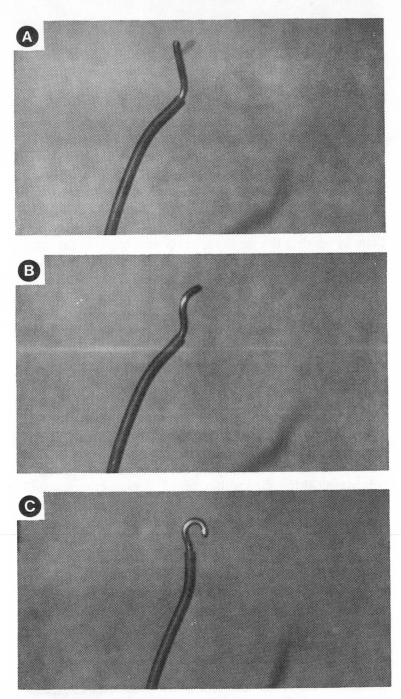

Fig. 3-16. Sequence for making an eye in a wire.

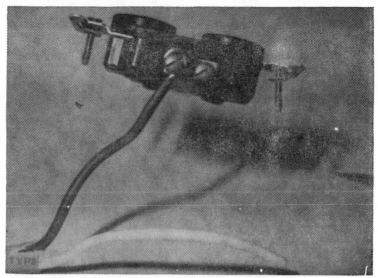

Fig. 3-17. Incorrect method of placing eye on a screw connector.

or ears are inserted under the screw head of the terminal, and the screw is then tightened for a sound electrical connection.

Other types of wire connectors are available, and directions of their proper use normally will be found on the carton in

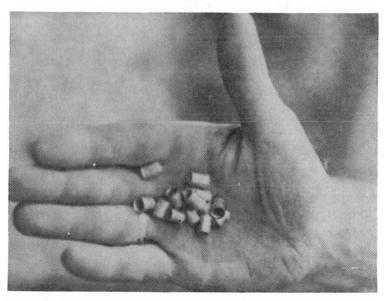

Fig. 3-18. One type of wire connectors.

which they are packaged. Some require special tools for proper use, but most may be used for electrical splices and connnections with conventional hand tools.

## SOLDERING

Although the soldering of electrical joints is seldom done in this day and age due to the variety of solderless lugs and wire connectors there may be times when the knowledge of soldering will come in handy around the farm or home. Or perhaps your preference will be to solder for a joint that will normally last a lifetime. In any case, the next few paragraphs will give basic instructions on methods used for proper soldering of electrical wires and joints.

To heat the wire and melt the solder a soldering gun or iron is used as described in Chapter 1. The tip of this device must be kept well cleaned, tinned, and heated for proper soldering. If the soldering tip is not hot enough, the solder will melt very slowly and become pasty, instead of flowing freely as it should.

It is generally agreed that solder of approximately forty percent tin and sixty percent lead is best for electrical work. It can be bought in the form of long bars, solid wire solder, and resin-core wire solder. The latter is very convenient for use by the homeowner as the resin carried in the hollow wire acts as a flux, automatically applied as the solder melts.

When soldering a wire splice place the heated soldering tip against (below) the wire splice to heat the wire; hold the tip so that as much area of the splice as possible comes into contact with it. Then melt a small drop of solder on the soldering tip by placing the solder wire in between the tip and the splice. This drop of solder should melt almost instantly and will provide a much greater area of metal-to-metal contact between the tip and the splice. This will cause the heat to flow into the splice many times faster, heating it thoroughly in a very few seconds.

Then, while still keeping the soldering tip hot and in contact with the bottom of the splice, run the solder wire along the top of the splice, allowing the melted solder to run down through the turns, until a good coating of solder covers the entire splice. However, never allow too much solder to

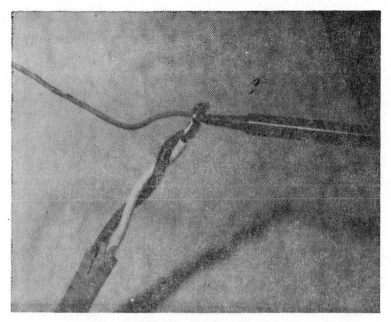

Fig. 3-19. Illustration showing the soldering of an electrical joint.

melt on the splice that will leave a large bulge; allow just enough so that a good coating remains on all turns of the splice. Figure 3-19 shows the soldering of an electrical joint.

## TAPING ELECTRICAL WIRE JOINTS

All soldered wire joints should be taped carefully to provide the same quality of insulation over the splice as over the rest of the wires.

Soldered joints in rubber-insulated wires should be covered with rubber and friction tape while joints in thermoplastic-insulated wires should be covered with pressure-sensitive thermoplastic adhesive tape, such as Scotch No. 33 for indoor use or Scotch No. 88 for outside use.

For covering rubber-insulated wires apply the rubber tape to the splice first to provide air and moisture tight insulation. The amount applied should be equal to the insulation that was removed. Then the friction tape is wrapped over the rubber tape to provide mechanical protection.

For joints in thermoplastic-insulated wires, just the one tape is necessary, that is, Scotch No. 33 or No. 88. To start

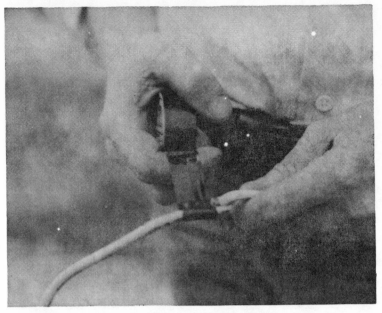

Fig. 3-20. Method of starting the taping of a splicer joint.

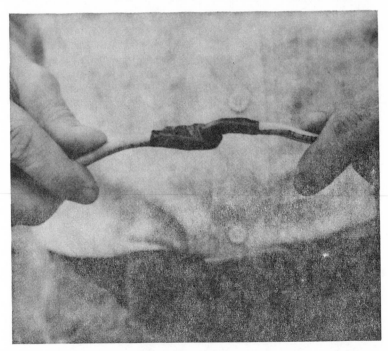

Fig. 3-21. A properly wrapped electrical joint.

46

the taping of a splice or joint, start the end of the tape at one end of the splice (see Fig. 3-20) slightly overlapping the insulation on the wires. Stretch it slightly while winding it on spirally. When the joint is completely covered with layers equal to the original insulation press or pinch the end of the tape down tightly onto the last turn to make it stick in place. A properly wrapped electrical joint is shown in Fig. 3-21.

# Chapter 4
# Electrical Materials
# for Home & Farm

The wiring method and materials on a given project in the home or around the farm are determined by several factors, namely, the requirements set forth in the National Electrical Code and Local Ordinances, the type of building construction, the location of the wiring in the building, the importance of the wiring system's appearance, and the relative costs of different wiring methods. In most instances, however, only two basic wiring methods will be used for home and farm electrical systems. They are sheathed cables of two or more conductors and what is termed as "raceway" systems.

In addition to these two basic wiring methods they may be further divided into two general divisions, which are open or exposed wiring and concealed wiring.

In open wiring systems, the outlets and cable or raceway systems are installed on the surfaces of the walls, ceilings, columns, etc. where they are in view and readily accessible. Such wiring is often used in basements, attics, barns, garages, and outbuildings, where appearance is not important and where it may often be desirable to make changes in the wiring.

Concealed wiring systems have all cable and raceway runs concealed inside of walls, partitions, ceiling, columns, and behind baseboards or molding where they are out of view and not readily accessible. This type of system is generally used in all new construction with finished interior walls, ceilings, floors, etc., and is the preferred type where a good appearance is important.

## CABLE SYSTEMS

There are four types of cable systems normally found in residential construction.

■ Nonmetallic sheath (NM) cable
■ Armored (BX) cable
■ Service-entrance (SE) cable
■ Underground-feeder (UF) cable

### Nonmetallic Sheath Cable

Nonmetallic sheathed cables are manufactured in two- and three-wire types with varying sizes of conductors. The jacket or covering consists of rubber or plastic. A typical example of NM cable is illustrated in Fig. 4-1. This type of cable may be concealed in the framework of residential type buildings or, in some instances, may be run exposed on the building surfaces.

### Armored Cable

Armored cable (called BX) is manufactured in two-, three- and four-wire assemblies, and with varying sizes of conductors. It is used in locations similar to those where type NM cable is used. The metallic spiral covering on BX cable offers a greater degree of mechanical protection than type NM cable, and the metal jacket also provides for a continuously grounded system without the need of additional grounding conductors. This type of cable may be used for underplas-

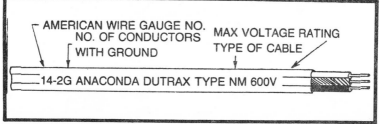

Fig. 4-1. Nonmetallic (NM) sheath cable.

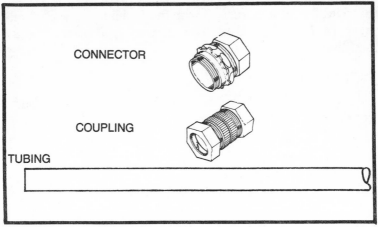

Fig. 4-2. EMT with its related couplings and connectors.

ter extentions as provided in the National Electrical Code, and embedded in plaster finish, brick, or other masonry, except in damp or wet locations. It also may be run or fished in the air voids of masonry block or tile walls except where such walls are exposed or subject to excessive moisture or dampness or are below grade.

## Service-Entrance Cable

Type SE (service-entrance) cable, when used for an electrical service must be installed as required in the National Electrical Code. This type of cable may be used in interior wiring sytems provided all the circuit conductors of the cable are insulated with rubber or thermoplastic insulation. Service-entrance cables without individual insulation on the grounded circuit conductor cannot be used as a branch circuit or as a feeder within a building, except when the cable has a final nonmetallic outer covering and when supplied by alternating current not exceeding 150 volts to ground. It may be used:

- As a branch circuit to supply only an electric range, wall-mounted oven, counter-mounted cooking unit or clothes dryer
- As a feeder to supply only other buildings on the same premises. It cannot be used for a feeder terminating within the same building in which it originates

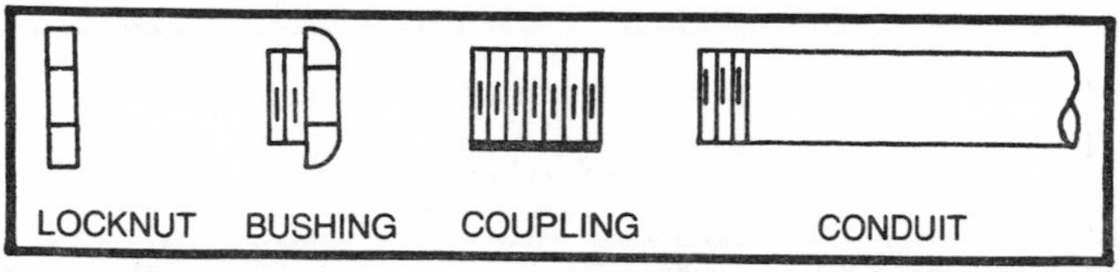

LOCKNUT   BUSHING   COUPLING        CONDUIT

Fig. 4-3. Rigid conduit with its related coupling.

## Underground-Feeder Cable

Type UF (underground-feeder) cable may be used underground, including direct burial in the earth, as a feeder or branch-circuit cable when provided with overcurrent protection (fused) at the rated ampacity as required by the National Electrical Code. When type UF cable is used above grade where it will come in direct contact with the rays of the sun its outer covering must be of the sun-resistant type.

## RACEWAY SYSTEMS

A raceway wiring system consists of an electrical wiring system in which two or more individual conductors are pulled into a conduit (pipe) or similar housing for the conductors after the raceway system has been completely installed. The basic raceways are rigid steel conduit, electrical metallic tubing (EMT), and PVC (polyvinylchloride) plastic. Figure 4-2 shows EMT with its related couplings and connectors, while Fig. 4-3 shows rigid conduit with its related coupling.

Other raceways include surface metal moldings (Fig. 4-4) and flexible metallic conduit (Fig. 4-5).

These raceways are available in standardized sizes and serves primarily to provide mechanical protection for the wires run inside and, in the case of metallic raceways, to provide a continuously grounded system.

Metallic raceways, properly installed, provide the greatest degree of mechanical and grounding protection, and provide maximum protection against fire hazards for the elec-

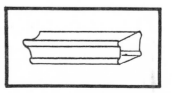

Fig. 4-4. Surface metal raceway.

Fig. 4-5. Flexible metallic conduit.

trical system. However, metallic raceways are more expensive to install, and for this reason, cable systems are normally used the most in home and farm electrical systems.

## USE OF DIFFERENT WIRING SYSTEMS

The selection of the proper wiring method is mainly determined from two factors. One is the use of special material required by different types of building construction. The other is from the different uses for which the electrical system is designed.

As an example, a homeowner may desire to install several lighting outlets and related wiring on a solid wood ceiling with exposed laminated beams. Since appearance is very important in this area, the wiring should preferably be concealed. However, due to the solid wood ceiling consisting of 2-inch thick tongue-and-groove boards with a built-up roof on top, concealing the wiring is very difficult to do. Still, a neat wiring installation can be had by using surface metal raceway such as shown in Fig. 4-6.

Another example might include a basement in a home where type NM or BX cable is used for wiring in the exposed wood ceiling joists; however, EMT or surface metal raceway would probably be used to feed the surface-mounted receptacles and wall switches on the masonry basement wall.

A service-entrance mast (Fig. 4-7), extending from a meter base located on carport through the roof in order for the power company's service drop to be connected, will almost always be installed in rigid metal conduit, while the load-side wires extending from the meter to the main electric switch or panel could either be installed in conduit (raceway system) or else consist of SE cable. Another ground wire, consisting of bare copper, will be used to ground the electric service to a cold water pipe entering the house.

Overhead wiring supplying electricity to various buildings around the farm can be installed with single conductors secured to insulators either on a pole or mounted on buildings. Or two- and three-conductor cables of the proper type could be used.

Outside underground wiring can be run in raceways (PVC or rigid conduit) or can be buried directly in the ground provided certain code regulations are followed.

Most of the wiring methods covered in this chapter can be installed with conventional hand tools such as the ones described in Chapter 1. If factory-made bends, threadless couplings and special fittings are used, even conduit systems can be installed with only hand tools. However, if the installation of EMT or rigid conduit is anticipated, the user will find that an inexpensive hand bender and the knowledge of its use will come in handy when kicks, offsets and saddles are required in the conduit. If you purchase a new EMT or conduit bender most of them will have complete instructions packed with them that tells exactly how to bend conduit to practically any shape or dimension. However, the basic requirements of conduit bending and installation follow to give you an idea of what the job entails.

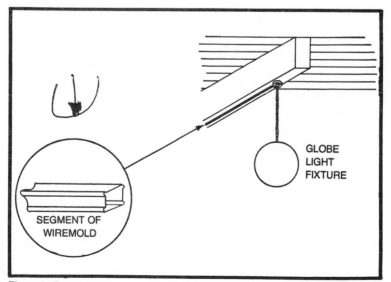

GLOBE
LIGHT
FIXTURE

SEGMENT OF
WIREMOLD

Fig. 4-6. Example of a new wiring installation using surface metal raceway.

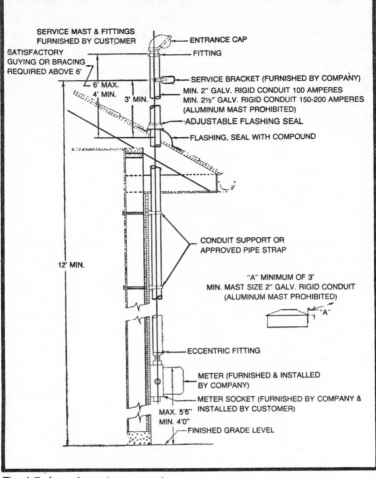

Fig. 4-7. A service-entrance mast.

## USING EMT

There are special tubing cutters on the market to cut EMT, but most of these cutters leave a large burr, and often a definite hump, inside the conduit. Therefore, most experienced electricians prefer to use a regular hacksaw with 32-teeth-per-inch blades when cutting EMT. If the cut is made square, only a small burr occurs on the inside which can quickly be reamed out with a special reamer; the end of a pair of side cutting pliers or a square shank screwdriver will also do for reaming most sizes of EMT.

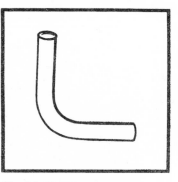

Fig. 4-8. EMT offset connector.　　Fig. 4-9. EMT factory elbow.

The usual installation of EMT requires many changes of direction in the runs, ranging from simple offsets where the conduit enters outlet or panel boxes to more complicated angular offset and saddles to miss various obstacles encountered in the run. However, if you find the bending of conduit impractical for you at this time there are various fittings available to get around any hand bending at all. For example, an EMT offset connector (Fig. 4-8) can be used at box terminations. Factory elbows (Fig. 4-9) may be used to make 90-degree bends. Conduit boxes (Fig. 4-10) may be used for any number of changes in direction of the conduit run without the need of bending the EMT with a hand bender. A junction box (Fig. 4-11) may also be used to make a 90-degree bend; merely run one conduit into the side of the box and begin the next piece out of the bottom.

To join ends of EMT together use EMT couplings (Fig. 4-12) of the proper type, that is, concrete-tight, weather-tight, etc. When terminating the ends of EMT into outlet boxes you will need an EMT connector as shown in Fig. 4-13.

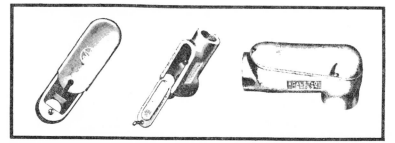

Fig. 4-10. Conduit boxes.

55

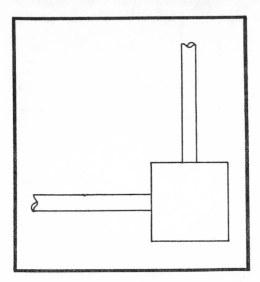

Fig. 4-11. A junction box.

If you want to try your hand at bending EMT remember that it's going to take practice, lots of practice. Use a roll type bender as shown in Fig. 4-14; this type of bender has high supporting sidewalls to prevent flattening or kinking of the tubing and a long arc that permits your making 90-degree bends in a single sweep.

Bends at 90 degrees are made by placing the EMT on the floor. Next measure and mark the conduit at the outer edge of the bend. Insert the bender with its pointer at the mark you made on the conduit. Then with your foot placed firmly on the bender make a 90-degree bend.

A saddle is made by first bending the EMT at a 45-degree angle (Fig. 4-15) at the point which will be directly over the obstruction. Next, determine the length of the straight legs

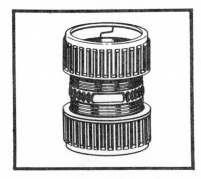

Fig. 4-12. Conventional EMT coupling.

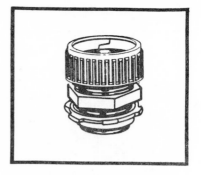

Fig. 4-13. EMT connector.

required to provide the necessary saddle rise. Complete the saddle by making a 22.5-degree bend at the measured point on each straight leg and in a direction opposite to the original 45-degree bend as shown in Fig. 4-16.

## OUTLET BOXES

The type of outlet box selected for a given job must be of a type for use with the wiring system you are installing (cable, EMT, conduit, etc.) and must be large enough to accommodate the number of wires which must be spliced or fed through the box.

The National Electrical Code (NEC) stipulates that outlet boxes shall be of sufficient size to provide free space for all conductors. Table 4-1 gives the trade sizes of the most common outlet boxes along with their dimensions and the number of conductors allowed in each.

Notice that Table 4-2 gives only the number of wires *of the same size* allowed in each box. There will be times when you will need to use a combination of different sizes in a single box. In this case Table 4-3 should be used in sizing the outlet box.

Fig. 4-14. EMT roll-type bender.

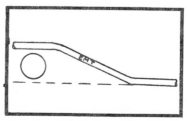

Fig. 4-15. First phase of bending a saddle in EMT.

Table 4-1. Outlet Box Specs & Number of Max Wires in Each

| Box (in.) | Capacity (cu. in.) | Max Number of Conductors | | | |
|---|---|---|---|---|---|
| | | AWG #14 | #12 | #10 | #8 |
| 3¼ × 1½ octagonal | 10.9 | 5 | 4 | 4 | 3 |
| 3½ × 1½ | 11.9 | 5 | 5 | 4 | 3 |
| 4 × 1½ octagonal | 17.1 | 8 | 7 | 6 | 5 |
| 4 × 2⅛ octagonal | 23.6 | 11 | 10 | 9 | 7 |
| 4 × 1½ square | 22.6 | 11 | 10 | 9 | 7 |
| 4 × 2⅛ square | 31.9 | 15 | 14 | 12 | 10 |
| 4 11/16 × 1½ square | 32.2 | 16 | 14 | 12 | 10 |
| 4 11/16 × 2⅛ square | 46.4 | 23 | 20 | 18 | 15 |
| 3 × 2 × 1½ device | 7.9 | 3 | 3 | 3 | 2 |
| 3 × 2 × 2 device | 10.7 | 5 | 4 | 4 | 3 |
| 3 × 2 × 2¼ device | 11.3 | 5 | 5 | 4 | 3 |
| 3 × 2 × 2½ device | 13 | 6 | 5 | 5 | 4 |
| 3 × 2 × 2¾ device | 14.6 | 7 | 6 | 5 | 4 |
| 3 × 2 × 3½ device | 18.3 | 9 | 8 | 7 | 6 |
| 4 × 2⅛ × 1½ device | 11.1 | 5 | 4 | 4 | 3 |
| 4 × 2⅛ × 1⅞ device | 13.9 | 6 | 6 | 5 | 4 |
| 4 × 2⅛ × 2⅛ device | 15.6 | 7 | 6 | 6 | 5 |

## Table 4-2. Outlet Box Selector Guide

### COMBINATIONS OF #14 AND #12 WIRES

| No. of Wires In Box W/O Ground Wire(s) | CUBIC-INCHES NEEDED | | | No. of Wires In Box W/O Ground Wire(s) | CUBIC-INCHES NEEDED | | |
|---|---|---|---|---|---|---|---|
| | CLAMPS ONLY | CLAMPS & 1 DEV. | CLAMPS & 2 DEVS. | | CLAMPS ONLY | CLAMPS & 1 DEV. | CLAMPS & 2 DEVS. |
| 2 #14 & 2 #12 | 13.00 | 15.25 | 17.50 | 6 #14 & 9 #12 | 36.75 | 39.00 | 41.25 |
| 2 #14 & 3 #12 | 15.25 | 17.50 | 19.75 | 7 #14 & 2 #12 | 23.00 | 25.25 | 27.50 |
| 2 #14 & 4 #12 | 17.50 | 19.75 | 22.00 | 7 #14 & 3 #12 | 25.25 | 27.50 | 29.75 |
| 2 #14 & 5 #12 | 19.75 | 22.00 | 24.25 | 7 #14 & 4 #12 | 27.50 | 29.75 | 32.00 |
| 2 #14 & 6 #12 | 22.00 | 24.25 | 26.50 | 7 #14 & 5 #12 | 29.75 | 32.00 | 34.25 |
| 2 #14 & 7 #12 | 24.25 | 26.50 | 28.75 | 7 #14 & 6 #12 | 32.00 | 34.25 | 36.50 |
| 2 #14 & 8 #12 | 26.50 | 28.75 | 31.00 | 7 #14 & 7 #12 | 34.25 | 36.50 | 38.75 |
| 2 #14 & 9 #12 | 28.75 | 31.00 | 33.25 | 7 #14 & 8 #12 | 36.50 | 38.75 | 41.00 |
| 3 #14 & 2 #12 | 15.00 | 17.25 | 19.50 | 7 #14 & 9 #12 | 38.75 | 41.00 | 43.25 |
| 3 #14 & 3 #12 | 17.25 | 19.50 | 21.75 | 8 #14 & 2 #12 | 25.00 | 27.25 | 29.50 |
| 3 #14 & 4 #12 | 19.50 | 21.75 | 24.00 | 8 #14 & 3 #12 | 27.25 | 29.50 | 31.75 |
| 3 #14 & 5 #12 | 21.75 | 24.00 | 26.25 | 8 #14 & 4 #12 | 29.50 | 31.75 | 34.00 |
| 3 #14 & 6 #12 | 24.00 | 26.25 | 28.50 | 8 #14 & 5 #12 | 31.75 | 34.00 | 36.25 |
| 3 #14 & 7 #12 | 26.25 | 28.50 | 30.75 | 8 #14 & 6 #12 | 34.00 | 36.25 | 38.50 |
| 3 #14 & 8 #12 | 28.50 | 30.75 | 33.00 | 8 #14 & 7 #12 | 36.25 | 38.50 | 40.75 |
| 3 #14 & 9 #12 | 30.75 | 33.00 | 35.25 | 8 #14 & 8 #12 | 38.50 | 40.75 | 43.00 |
| 4 #14 & 2 #12 | 17.00 | 19.25 | 21.50 | 9 #14 & 2 #12 | 27.00 | 29.25 | 31.50 |
| 4 #14 & 3 #12 | 19.25 | 21.50 | 23.75 | 9 #14 & 3 #12 | 29.25 | 31.50 | 33.75 |
| 4 #14 & 4 #12 | 21.50 | 23.75 | 26.00 | 9 #14 & 4 #12 | 31.50 | 33.75 | 36.00 |
| 4 #14 & 5 #12 | 23.75 | 26.00 | 28.25 | 9 #14 & 5 #12 | 33.75 | 36.00 | 38.25 |
| 4 #14 & 6 #12 | 26.00 | 28.25 | 30.50 | 9 #14 & 6 #12 | 36.00 | 38.25 | 40.50 |
| 4 #14 & 7 #12 | 28.25 | 30.50 | 32.75 | 9 #14 & 7 #12 | 38.25 | 40.50 | 42.75 |
| 4 #14 & 8 #12 | 30.50 | 32.75 | 35.00 | 9 #14 & 8 #12 | 40.50 | 42.75 | 45.00 |
| 4 #14 & 9 #12 | 32.75 | 35.00 | 37.25 | 10 #14 & 2 #12 | 29.00 | 31.25 | 33.50 |
| 5 #14 & 2 #12 | 19.00 | 21.25 | 23.50 | 10 #14 & 3 #12 | 31.25 | 33.50 | 35.75 |
| 5 #14 & 3 #12 | 21.25 | 23.50 | 25.75 | 10 #14 & 4 #12 | 33.50 | 35.75 | 38.00 |
| 5 #14 & 4 #12 | 23.50 | 25.75 | 28.00 | 10 #14 & 5 #12 | 35.75 | 38.00 | 40.25 |
| 5 #14 & 5 #12 | 25.75 | 28.00 | 30.25 | 10 #14 & 6 #12 | 38.00 | 40.25 | 42.50 |
| 5 #14 & 6 #12 | 28.00 | 30.25 | 32.50 | 10 #14 & 7 #12 | 40.25 | 42.50 | 44.75 |
| 5 #14 & 7 #12 | 30.25 | 32.50 | 34.75 | 11 #14 & 2 #12 | 31.00 | 33.25 | 35.50 |
| 5 #14 & 8 #12 | 32.50 | 34.75 | 37.00 | 11 #14 & 3 #12 | 33.25 | 35.50 | 37.75 |
| 5 #14 & 9 #12 | 34.75 | 37.00 | 39.25 | 11 #14 & 4 #12 | 35.50 | 37.75 | 40.00 |
| 6 #14 & 2 #12 | 21.00 | 23.25 | 25.50 | 11 #14 & 5 #12 | 37.75 | 40.00 | 42.25 |
| 6 #14 & 3 #12 | 23.25 | 25.50 | 27.75 | 11 #14 & 6 #12 | 40.00 | 42.25 | 44.50 |
| 6 #14 & 4 #12 | 25.50 | 27.75 | 30.00 | 12 #14 & 2 #12 | 33.00 | 35.25 | 37.50 |
| 6 #14 & 5 #12 | 27.75 | 30.00 | 32.25 | 12 #14 & 3 #12 | 35.25 | 37.50 | 39.75 |
| 6 #14 & 6 #12 | 30.00 | 32.25 | 34.50 | 12 #14 & 4 #12 | 37.50 | 39.75 | 42.00 |
| 6 #14 & 7 #12 | 32.25 | 34.50 | 36.75 | 12 #14 & 5 #12 | 39.75 | 42.00 | 44.50 |
| 6 #14 & 8 #12 | 34.50 | 36.75 | 39.00 | | | | |

### COMBINATIONS OF #12 AND #10 WIRES

| No. of Wires In Box W/O Ground Wire(s) | CLAMPS ONLY | CLAMPS & 1 DEV. | CLAMPS & 2 DEVS. | No. of Wires In Box W/O Ground Wire(s) | CLAMPS ONLY | CLAMPS & 1 DEV. | CLAMPS & 2 DEVS. |
|---|---|---|---|---|---|---|---|
| 2 #12 & 2 #10 | 14.50 | 17.00 | 19.50 | 6 #12 & 2 #10 | 23.50 | 26.00 | 28.50 |
| 2 #12 & 3 #10 | 17.00 | 19.50 | 22.00 | 6 #12 & 3 #10 | 26.00 | 28.50 | 31.00 |
| 2 #12 & 4 #10 | 19.50 | 22.00 | 24.50 | 6 #12 & 4 #10 | 28.50 | 31.00 | 33.50 |
| 2 #12 & 5 #10 | 22.00 | 24.50 | 27.00 | 6 #12 & 5 #10 | 31.00 | 33.50 | 36.00 |
| 2 #12 & 6 #10 | 24.50 | 27.00 | 29.50 | 6 #12 & 6 #10 | 33.50 | 36.00 | 38.50 |
| 2 #12 & 7 #10 | 27.00 | 29.50 | 32.00 | 6 #12 & 7 #10 | 36.00 | 38.50 | 41.00 |
| 2 #12 & 8 #10 | 29.50 | 32.00 | 34.50 | 6 #12 & 8 #10 | 38.50 | 41.00 | 43.50 |
| 2 #12 & 9 #10 | 32.00 | 34.50 | 37.00 | 7 #12 & 2 #10 | 25.75 | 28.25 | 30.75 |
| 3 #12 & 2 #10 | 16.75 | 19.25 | 21.75 | 7 #12 & 3 #10 | 28.25 | 30.75 | 33.25 |
| 3 #12 & 3 #10 | 19.25 | 21.75 | 24.25 | 7 #12 & 4 #10 | 30.75 | 33.25 | 35.75 |
| 3 #12 & 4 #10 | 21.75 | 24.25 | 26.75 | 7 #12 & 5 #10 | 33.25 | 35.75 | 38.25 |
| 3 #12 & 5 #10 | 24.25 | 26.75 | 29.25 | 7 #12 & 6 #10 | 35.75 | 38.25 | 40.75 |
| 3 #12 & 6 #10 | 26.75 | 29.25 | 31.75 | 7 #12 & 7 #10 | 38.25 | 40.75 | 43.25 |
| 3 #12 & 7 #10 | 29.25 | 31.75 | 34.25 | 8 #12 & 2 #10 | 28.00 | 30.50 | 33.00 |
| 3 #12 & 8 #10 | 31.75 | 34.25 | 36.75 | 8 #12 & 3 #10 | 30.50 | 33.00 | 35.50 |
| 3 #12 & 9 #10 | 34.25 | 36.75 | 39.25 | 8 #12 & 4 #10 | 33.00 | 35.50 | 38.00 |
| 4 #12 & 2 #10 | 19.00 | 21.50 | 24.00 | 8 #12 & 5 #10 | 35.50 | 38.00 | 40.50 |
| 4 #12 & 3 #10 | 21.50 | 24.00 | 26.50 | 8 #12 & 6 #10 | 38.00 | 40.50 | 43.00 |
| 4 #12 & 4 #10 | 24.00 | 26.50 | 29.00 | 9 #12 & 2 #10 | 30.25 | 32.75 | 35.25 |
| 4 #12 & 5 #10 | 26.50 | 29.00 | 31.50 | 9 #12 & 3 #10 | 32.75 | 35.25 | 37.75 |
| 4 #12 & 6 #10 | 29.00 | 31.50 | 34.00 | 9 #12 & 4 #10 | 35.25 | 37.75 | 40.25 |
| 4 #12 & 7 #10 | 31.50 | 34.00 | 36.50 | 9 #12 & 5 #10 | 37.75 | 40.25 | 42.75 |
| 4 #12 & 8 #10 | 34.00 | 36.50 | 39.00 | 10 #12 & 2 #10 | 32.50 | 35.00 | 37.50 |
| 4 #12 & 9 #10 | 36.50 | 39.00 | 41.50 | 10 #12 & 3 #10 | 35.00 | 37.50 | 40.00 |
| 5 #12 & 2 #10 | 21.25 | 23.75 | 26.25 | 10 #12 & 4 #10 | 37.50 | 40.00 | 42.50 |
| 5 #12 & 3 #10 | 23.75 | 26.25 | 28.75 | 10 #12 & 5 #10 | 40.00 | 42.50 | 45.00 |
| 5 #12 & 4 #10 | 26.25 | 28.75 | 31.25 | 11 #12 & 2 #10 | 34.75 | 37.25 | 39.75 |
| 5 #12 & 5 #10 | 28.75 | 31.25 | 33.75 | 11 #12 & 3 #10 | 37.25 | 39.75 | 42.25 |
| 5 #12 & 6 #10 | 31.25 | 33.75 | 36.25 | 11 #12 & 4 #10 | 39.75 | 42.25 | 44.75 |
| 5 #12 & 7 #10 | 33.75 | 36.25 | 38.75 | 12 #12 & 2 #10 | 37.00 | 39.50 | 42.00 |
| 5 #12 & 8 #10 | 36.25 | 38.75 | 41.25 | | | | |
| 5 #12 & 9 #10 | 38.75 | 41.25 | 44.00 | | | | |

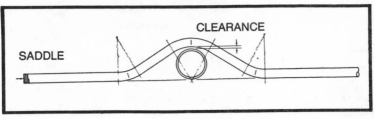
Fig. 4-16. Final phase of making a saddle in EMT.

If you were faced with the problem of selecting an outlet box on a conduit system in which four AWG #12, two AWG #10 and two AWG #8 wires will enter you can use the following method. Since this is a conduit system the box will contain no clamps, devices, ground wires, etc.

Two AWG #12 wires will have a volume of 4.5 (2.25 + 2.25) cubic inches; two AWG #10 wires will have a volume of 5 (2.50 + 2.50) cubic inches; two AGW #8 wires will have a total volume of 6 (3.00 + 3.00) cubic inches. Adding these figures together, we have a total of 15.5 cubic inches required

Table 4-3. Volume in Outlet Box Required by Conductor

| AWG | CU IN. |
|-----|--------|
| 14  | 2.00   |
| 12  | 2.25   |
| 10  | 2.50   |
| 8   | 3.00   |
| 6   | 5.00   |

for all the wires. Now referring to Table 4-1 in we see that a 4-inch octagonal box, 1½ inches deep, has a capacity of 17.1 cubic inches and would suffice for our application.

The outlet box selector guide in Table 4-2 is also very helpful for quickly selecting the proper size box for any given combination of wires and is recommended for use by the homeowner.

# Chapter 5
# Troubleshooting
# the Home Electrical System

The homeowner can save a great deal of time and money by doing his own troubleshooting when a problem occurs in the home electrical system. Even if an electrical contractor or repairman has to be called in to correct the problem the homeowner will still be money ahead by knowing the probable cause when the repairman arrives.

Troubleshooting covers a wide range of electrical problems from finding a short circuit in an appliance cord to tracing out troubles in complex control circuits used around the home. However, in nearly all cases, the homeowner can determine the cause of the trouble by using an inexpensive testing instrument and going about locating the troubles in a systematic and methodical manner, testing one part of the circuit or system at a time, until the trouble is located.

Always keep in mind that every electrical problem occurring around the home can solved. It is the purpose of this chapter to show the reader exactly how to go about solving the more conventional residential electrical problems in a safe and logical manner.

## TYPES OF ELECTRICAL FAULTS

In general, there are only three basic electrical faults, namely, a short circuit, an open circuit, and a change in

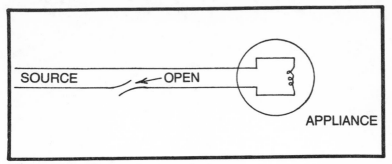

Fig. 5-1. Wiring diagram of an open circuit.

electrical value. To illustrate let's assume that several lights in the home suddenly stop burning. The most probable cause of this fault would be an open circuit caused by a blown fuse or tripped circuit breaker. Either one would open the circuit to which the lights were attached. Since the circuit is open no current will flow in the circuit, as shown in Fig. 5-1, and the lights would not burn. If, upon examination, the circuit breaker is tripped then some fault in the circuit caused the circuit breaker to trip in most cases. The faults could include a loose wire at the panelboard causing the wire to heat and trip the breaker, a short circuit developed by perhaps worn insulation on the wires or perhaps a change in electrical value by plugging in a heavy appliance on the circuit which caused the circuit to be overloaded.

If the trouble was found to be a loose wire at the screw terminal on the circuit breaker remove the wire entirely from the terminal to clean it because the excessive heat and arcing probably formed a coating on the bare wire. Next, insert the wire under the screw terminal, then tighten solidly. This should solve the problem.

Should the problem be due to a short circuit, which is an undesired current path that permits the electrical current to bypass the load, the short must be found and corrected. Methods of locating and correcting short circuits will be discussed later in this chapter.

If the cause of the tripped circuit breaker was due to an overloaded circuit then obviously the load on the circuit must be lightened by disconnecting one or more devices plugged into the circuit.

## Fuses

On circuits protected by plug fuses the nature of the problem can often be determined by the appearance of the fuse window. For example, if the window is clear and the metal strip appears to be intact, the fuse is probably not blown and the problem lies elsewhere in the circuit. However, it is best to check the fuse to be certain.

The best way to check a plug fuse is to unscrew it from its socket and connect the leads of an ohmmeter or continuity tester as shown in Fig. 5-2. If the fuse is good the pointer of the ohmmeter will swing all away across the scale to zero; if a continuity test lamp is used the lamp will burn. If the fuse is bad, the pointer will not move at all nor will the test lamp burn.

Figure 5-3 shows how to test a plug fuse with a voltmeter, or voltage test lamp, should an ohmmeter not be available. Place one lead of the test lamp on the neutral block in the panelboard or fuse cabinet and the other on the load side (usually the screw terminal where the hot circuit wire is connected) of the fuse. If a full reading is obtained (110 to 120 volts) on the voltmeter or the test lamp lights to full brilliance then the fuse is good. If the meter does not show a reading or the test lamp does not light at all, then the fuse is more than likely blown and the window of the fuse should be examined very closely as this will usually give some indication of what caused the fuse to blow.

If the window is clear but you notice that the metal strip inside of the window is broken it was probably a light overload

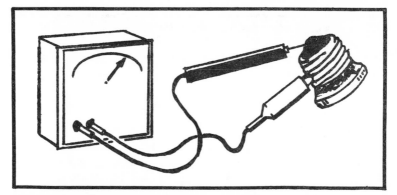

Fig. 5-2. Method of testing plug fuse with ohmmeter.

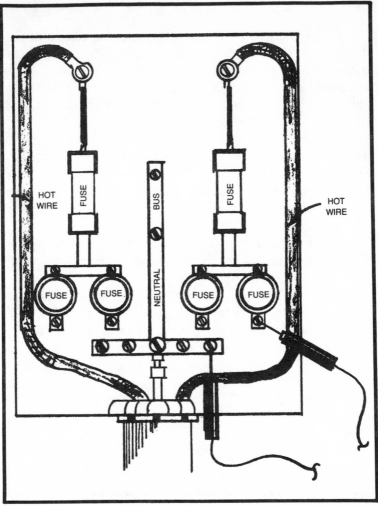

Fig. 5-3. Method of testing plug fuse with a voltmeter.

that caused the fuse to blow. Perhaps an iron or toaster was plugged in on the circuit. In any event, always check what caused the overload before replacing the fuse with a new one.

A badly blackened fuse window normally indicates a violent blowing of the fuse such as a severe overload or short circuit. Again, always check all possible causes and conditions in the circuit before replacing the fuse. In most cases you will find that someone in the house has just connected a heavy appliance on the circuit or perhaps someone stumbled over a lamp cord causing the wires connected to the lamp to be jerked loose and touch each other, thus shorting the circuit.

## Short Circuits

A short circuit (Fig. 5-4) is probably the most common cause of electrical problems in the home or around the farm. Sometimes the short is between two wires due to faulty insulation or else between one hot wire and a grounded object.

In locating a short circuit all loads connected to the circuit should be disconnected one at a time until the fault is found. This can be done by unscrewing the fuse or place the circuit breaker handle to the off position to insure that no line current will flow in the circuit. Then attach one lead of an ohmmeter to the load side of the circuit and the other to the disconnected neutral conductor. Set the ohmmeter to the lowest ohms scale. All appliances should be plugged in but switched off. The pointer of the needle should swing to some point from halfway to zero on the scale, which indicates a short in the wiring. Have someone unplug all of the electrical devices connected to the circuit until the pointer returns to its resting point (no reading); this is the appliance that is giving the trouble. A great many short circuits traced in this manner will be found in defective cords of portable appliances and other electrical devices.

If removing the various devices on the circuit does not correct the short circuit then it must be in the circuit wiring itself. When this occurs begin at the panelboard or fuse cabinet and work along the circuit wiring, opening up the various outlet boxes on the circuit and examining the wiring in

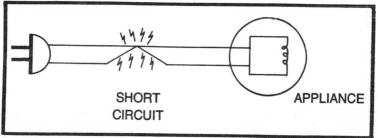

Fig. 5-4. Wiring diagram of a short circuit in an appliance cord.

each and making a test at each. In doing so all connections, splices, etc. must be opened.

To illustrate suppose that the first outlet on the circuit from the panelboard was a basement lighting outlet. With the circuit breaker open or the fuse unscrewed the fixture is removed. Then the splices in the outlet box are unwound and loosened. If the ohmmeter shows no reading across the wires from the fixture back to the panelboard, but the meter's pointer swings to zero between the two wires leading out of the outlet box to another outlet, then we can assume that the fault lies further along in the circuit.

Next the splices in the following outlet box are taken apart to make another test with the ohmmeter. This time the meter indicates that the circuit wires from the second outlet back to the first outlet show a short, while the wires leaving this second outlet and going to the next one are clear. Therefore the fault lies somewhere between the first and second outlet. Now the going is easier because the fault is pinpointed to a given area.

In most cases of this sort, short circuits in the wiring system will occur at the outlet boxes where perhaps a wire has vibrated loose from a terminal on a receptacle, switch, or other wiring device. It has shorted the circuit by coming into contact with another base wire or against the grounded outlet box. At times, however, the trouble may be located between the outlet boxes, especially if home repairs or remodeling has just taken place. If so the short circuit could be caused by a nail being driven through a piece of NM cable, or maybe the circuit wires were accidentally cut during the remodeling. There are any number of possible causes, but all can be located quickly if the circuit is traced and the trouble is pinpointed as described previously.

### Loose Neutral Wire

Whenever some of the lights on the home electrical system are found to be burning excessively bright while others are burning very dimly, the likely cause is a loose neutral wire somewhere in the system and especially at the main switch or panelboard. If this problem occurs check and tighten all neutral conductors on the panel board or main switch neutral block terminals, including the service-entrance neutral wire feeding the block.

## Partial Ground Fault

A partial ground on a circuit is not only troublesome but can needlessly run your electric bill up so that you are paying for current you are not using. The majority of these partial grounds will not trip the circuit breaker or blow a fuse and the fault will therefore continue for long periods of time without the homeowner noticing it, except for a high electric bill.

If your electric bill does go up in cost without any good explanation you probably have a partial ground on one of the house circuits. Even if you haven't noticed any great increase in your electric bill it wouldn't hurt anything to make periodic checks for partial grounds about once each year.

First, leave every appliance, lamp and any other normally operated electrical device plugged into their respective outlet, but make certain that none or operating. Then look at the dial on your electric meter to see if it is turning. If you are certain that every electric device in your home (including electric clocks) is not operating and yet the dial on the electric meter still turns, then there is certain to be a partial ground on one of the electrical devices.

To find which electrical device has the partial ground, have someone watch the dial on the electric meter as you begin unplugging and turning off the various devices one at a time. When the dial stops, you have found the faulty device.

In the majority of ground faults or partial grounds found in the home electric water heaters and electric pump motors seem to be the main appliances developing the trouble. For this reason these appliances should be checked first if the initial tests indicate a fault.

The simple test to determine a partial ground will more than likely be sufficient for most home electrical systems, but there are some instances where this test cannot be used due to the location of the partial ground. If the amount of your electric bill indicates that it is higher than it should be and you can find no faulty appliances by the method indicated, then have the power company check their metering equipment. If

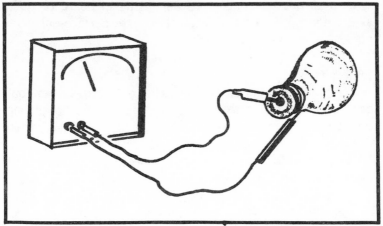

Fig. 5-5. Method of testing an incandescent lamp with ohmmeter.

the metering equipment checks out all right then your home electrical system should be checked by the power company or an electrical contractor with a megger instrument. Such an instrument will locate any partial ground fault immediately.

## TEST PROCEDURES

The following examples will show the reader how to test and repair electrical appliances and circuits in the home and around the farm. Every conceivable problem could not be included, but the examples given should be sufficient for most problems that normally occur.

### Electric Lamps

Electric lamps, both incandescent and fluorescent, may be checked with an ohmmeter or continuity test lamp. Figure 5-5 shows the position of the test leads for common incandescent lamps. One lead is held against the outer screw base while the other lead is placed against the small contact point on the bottom of the base. If the ohmmeter pointer remains at infinity ($\infty$) or the test lamp does not light the filament is burned out. However, if the test lamp lights or the ohmmeter pointer swings slightly on the scale to any degree towards zero the lamp is good.

Three-way lamps that have a special base and fit in a special socket may be checked in a similar manner except that the continuity check is made between the outer screw base and each of the inner contact points.

## Motor Operated Appliances

Small motor-operated appliances such as vacuum cleaners, fans, clocks and blenders may be checked with the ohmmeter by placing the instrument leads on the appliance plug as shown in Fig. 5-6. The smallest motors such as those used in electric clocks should show a reading of approximately 1000 ohms if the motor winding is good; around 50 ohms for motors in fans, vacuum cleaners, etc., and approximately 5 ohms for larger motors such as those used to drive washing machines, refrigerators and water pumps. Remember that the smaller the motor, the higher will be the resistance (in ohms) if the winding is good. Of course, the switch must be on when making such resistance tests.

## Heating Appliances

Tests on appliances with heating elements are made by attaching the cord to the appliance and then placing the in-

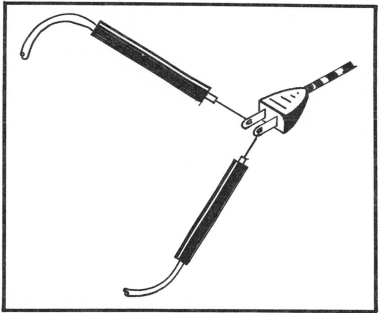

Fig. 5-6. Position of test leads on a plug for various tests.

strument or test light leads on the plug as shown in Fig. 5-6. Coffee makers, irons, heating pads, and similar electric appliances fall under this category.

An infinity (∞) reading indicates that either the cord or heating element is open; a zero reading indicates a short circuit. Make certain the switch or thermostat is working. A reading from 15 to 20 ohms indicates that the element is good.

If the reading indicates an open circuit remove the appliance cord from the device. Connect the two wires together on the end opposite from the plug. Then attach the instrument leads. An infinity (∞) reading indicates that the cord is open while a zero reading indicates that the wires are okay.

Methods of troubleshooting various other household and farm appliances, wiring devices, and electrical systems will be covered in other chapters under their respective headings. For example, troubleshooting overloaded branch circuits will be covered in Chapter 6, overloaded electric services will be covered in Chapter 10, etc.

# Chapter 6
# Adding Duplex Receptacles

For new construction such as the building of a new house or the addition of a new wing on an existing house, the National Electrical Code specifically states the minimum requirement for the location of receptacles in residential buildings:

> ...In every kitchen, family room, dining room..., receptacle outlets shall be installed so that no point along the floor line in any wall space is more than 6 feet, measured horizontally, from an outlet in that space, including any wall space 2 feet wide or greater and the wall space occupied by sliding panels in exterior walls... Receptacle outlets shall, insofar as practical, be spaced equal distances apart. Receptacle outlets in floors shall not be counted as part of the required number of receptacle outlets unless located close to the wall....

Basically, this states that any appliance, table lamp, etc. placed in any room in the residence and along the wall can be plugged in an outlet with only a 6-foot extension cord from the appliance; no additional extention cords would be necessary. If one of your electrical devices required an extension cord longer than 6 feet to operate the device anywhere in your home, your outlets are not spaced according to the National Electrical Code. However, if your home was built prior to 1960, don't let this worry you, as few houses built before 1960 have this many duplex receptacles.

Rather than begin a project to completely renovate your home's electrical system should it not come up to Code standards, let's look at the need for additional duplex receptacles from another angle. First, make a quick survey of all the rooms in your home. If you found that several extension cords are required to furnish all of your electrical appliances with power, then it's time to add additonal receptacles. Many of these cheap extention cords are not only fire hazards, but also make obstacles for you and your guests to trip over.

Are you finding that you are blowing fuses more often than you should, or do some of the lights go dim when a toaster or coffee maker is plugged in ? If so, then it's time to add additional receptacles and circuits.

Perhaps you have just purchased a new freezer or other appliance, but the area you want it located does not have an outlet to plug it into. Again, an additional receptacle is in order.

Due to undersized wire and overloaded circuits many home electrical systems cause energy losses—wasted watts—in the form of low voltage. In other words, if your electric service is rated at 120 volts at your main switch and if your circuits are overloaded, there may be only 100 to 105 volts by the time the current arrived at the appliance to be operated. Excessive voltage drop results in great losses to the owners by having appliances (especially motor-driven ones) operate at less than their normal efficiency; your monthly electric bill therefore includes charges for watts which you are not using. They are going through your meter, but they are expended in your overloaded circuits and not the electric appliances for which they are intended.

If you suspect that your electrical system is overloaded you will be money ahead by correcting the situation. One step could be to add additional duplex receptacles and new circuits of the correct wire size.

## PLAN AHEAD

In starting to install one or more duplex receptacles on your present electrical system there are certain general steps to be followed. Before cutting any openings in your walls or running any wires look the situation over very carefully. In doing so much material, labor and needless cutting and patching can usually be saved.

When deciding upon the location of an additional outlet try to locate it where electrical wiring can be installed easily. For example, the best location would be where the structure is open and readily available to receive the wiring such as in an unfinished garage, attic, or basements. However, this is not usually the case when additional outlets are installed in an existing residence; rather, they are most often installed on finished walls. Nevertheless try to locate the outlet so that the wires can be fished from an opening in the attic or basement to the new outlet by the easiest route.

The material used will depend upon the location of the outlet and construction of the area through which the wires must run. The following step-by-step methods for various installation situations should suffice to give the homeowner the necessary information to install duplex receptacles anywhere in the home or around the farm.

## CONCEALED WIRING IN ACCESSIBLE SPACES

The floor plan of an area in a home is shown in Fig. 6-1. Note that a new duplex receptacle is to be installed on a finished wall that is adjacent to an unfinished wall space.

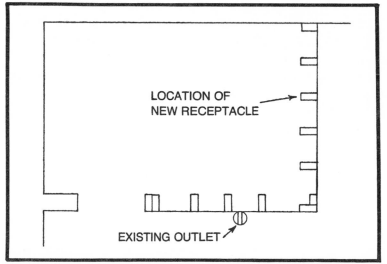

Fig. 6-1. Floor plan of an area in a home.

Fig. 6-2. Checking the height of an existing receptacle in an area near the one to be installed.

The first step is to remove the cover plate on one of the existing receptacles in the area to check the height of this outlet from the finished floor to the bottom of the outlet box as shown in Fig. 6-2. This measurement will be used to lay out the height of the new outlet.

Once the height of the new outlet is determined and the approximate location is marked on the wall make certain that the hole to be cut will clear any studs or other obstacles in the finished wall. In this case, the position of the studs can be checked visually by examining the unfinished attic space. In areas where this is not possible, the position of the studs can be found by light tapping on the wall surface; a dead sound indicates that something is behind the wall while a hollow sound indicates that the space is clear. A magnetic device may be purchase for less than one dollar which can detect nails driven into studs behind the wall if you have a tin ear.

After having exactly located the spot for the new outlet box place the box against the wall and mark the outline of it on the wall as shown in Fig. 6-3. Then with a wood bit drill two holes *inside* of this marked area in opposite corners as shown

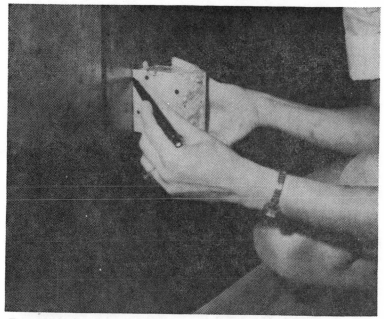

Fig. 6-3. Marking the outline of the new outlet with an outlet box.

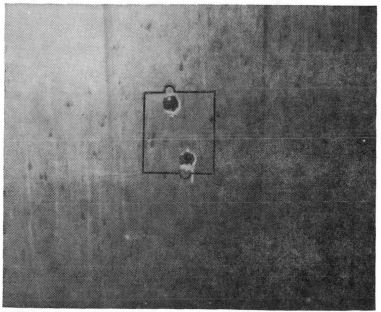

Fig. 6-4. Holes drilled inside of this marked area to allow for the insertion of a saw blade.

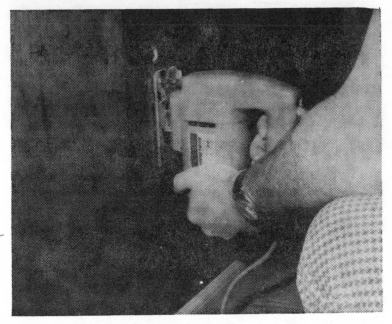

Fig. 6-5. Cutting the hole for a new receptacle.

in Fig. 6-4. The holes should be large enough to allow the blade of a keyhole or saber saw to start cutting out the opening, as shown in Fig. 6-5.

With a screwdriver (Fig. 6-6) remove one of the knock-outs in the outlet box by inserting the blade in the slot, bending the knockout outward and twisting until the small piece of metal holding the knockout snaps. Then insert the box in the opening until the face of the box is flush with the finished wall.

At this point it is necessary to secure the box. There are many different methods for this. The correct one will be dictated by the type of box used and the accessibility of the box. In this case since the other side of the wall is open we will use two 16-penny nails to secure the box to the stud on the opposite side from the finished wall. The secured box is shown in Fig. 6-7.

Referring again to the floor plan in Fig. 6-1 you will notice that an existing outlet is only a few feet away from the new one. Since the existing circuit is lightly loaded we will use this existing receptacle to feed the new one. However, before

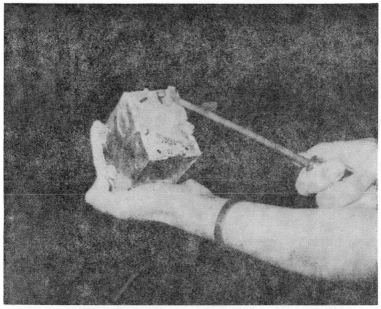

Fig. 6-6. Removing a knockout from an outlet box with a screwdriver.

Fig. 6-7. Illustration of an outlet box secured with 16-penny nails.

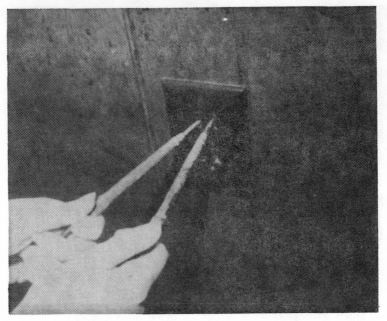

Fig. 6-8. Testing outlet for voltage.

doing another thing make absolutely certain that the power is turned off on this circuit. This is done by inserting the test leads into the outlet as shown in Fig. 6-8 while the outlet is hot. Your voltmeter or test lamp should indicate that the outlet is alive. Then, leaving the leads in the receptacle, turn off the circuit breaker or pull the fuse providing overcurrent protection for the circuit. If in doubt pull the main switch, but still check the outlet afterwards with your voltmeter or test lamp to insure that the outlet is dead.

Remove the two screws holding the duplex receptacle in the box as shown in Fig. 6-9. Pull the receptacle out of the box as far as it will go without removing the wires attached to it. Loosen the clamp inside of the box holding the wires. Remove one of the knockouts from the outlet box as was described previously. If NM cable is used, loosen the ground wire from the box or receptacle.

With a drill and wood bit, drill holes in the studs to allow a new piece of cable to be pulled from the existing outlet box to the new outlet location (Fig. 6-10). Pull the cable through the hole and allow about 1 foot excess on each end of the cable.

Fig. 6-9. Removing duplex receptacle from outlet box.

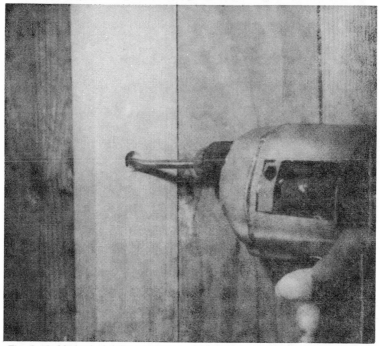

Fig. 6-10. Method of drilling holes in wall studs to accept electrical wiring.

79

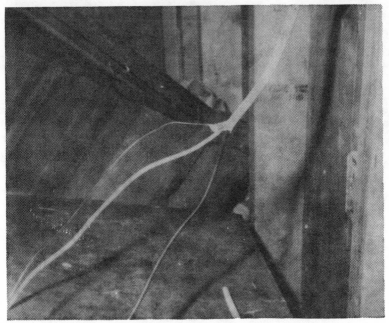

Fig. 6-11. Illustration of cable ready to be inserted in an outlet box.

Strip approximately 10 to 12 inches of cable sheathing from the cable as described in Chapter 3. Strip and clean about 1½ inches from the ends of each wire inside of the cable. The result should look like the cable in Fig. 6-11.

Insert one end of the cable through the knockout in the new outlet box and under the cable clamp so that the clamp fits over the cable sheathing, not the wires inside of the cable. Repeat this on the opposite end of the cable, that is, at the existing outlet (see Fig. 6-12).

In the existing outlet box wrap the bare ground wires around each other with sort of a modified rattailed splice. Insert one of the ground wires under a ground clip or else under an approved ground screw secured to the metal outlet box. Then place the other ground wire under the green grounding screw on the receptacle.

Make an eye in the bare wire with white insulation, and snug it under the light colored screw and tighten. Then do the same with the wire with black insulation, and place it under the brass colored screw and tighten. Replace the existing receptacle back in the outlet box, and install the cover plate.

Now repeat the previous step on the end where the new receptacle is connected. Before installing the receptacle or cover make sure that all connections are tight and that no insulation was damaged during the process. If you happened to nick the insulation slightly during the installation tape the nick with a small piece of electric tape.

Once everything is in place activate the circuit breaker or replace the fuse for the circuit. You now have the use of another duplex receptacle. Check both outlets again with your voltmeter or test lamp to ascertain that they are working. If you followed these procedures to the letter no trouble should develop.

If the circuit breaker trips or the fuse blows, stop. Make certain the circuit is turned off. Check your wiring for bare spots or nicked insulation. It is shorting somewhere.

This installation just described was relatively simple in that the unfinished attic allowed complete access to the outlets and for installing the cable. However, as mentioned previously this is not always the case. The wire may have to be fished down the wall partitions from an unfinished attic above

Fig. 6-12. Illustration of wires inserted in outlet box.

Fig. 6-13. One method of fishing cable in existing wall partitions from an unfinished attic above.

(Fig. 6-13) or from the basement below (Fig. 6-14). The basic procedures are practically the same for the actual installation as the previous example except that different types of box holders will have to be used. The outlet box in Fig. 6-15 is made for existing wiring and is fitted with a wallboard hanger attachment which will secure the box snugly to the wall without the need of additional anchors.

To install a wallboard hanger box cut the hole in the wall the exact size of the outlet box. Assemble box and hanger, but only partially tighten bolt holding the assembly. Insert cable into box and tighten cable clamp before pushing the entire assembly into the wall hole until the sides of the hanger spring free on the inside of the wall or partition. Tighten the assembly by tightening the bolt on the inside of the box. When tight pressure on both the inside and outside of the wall gives the box a rigid installation.

A number of methods have already been described for pulling and fishing cables into walls and openings in existing finished buildings. With a little ingenuity and careful planning you will be able to solve almost any problem encountered.

In pulling cables into spaces between the joists in ceilings or floors or between studs in walls, a flashlight or drop light placed in or near the outlet box hole is often a great help in fishing the wires in, or when catching them with a hook to pull them out of the outlet opening.

Where it is necessary to remove floorboards, baseboards, molding, door and window trim, etc. to route wiring to outlets in the home (see Fig. 6-16) it should be done

Fig. 6-14. Method of fishing cable in wall partition from basement below.

Fig. 6-15. An outlet box designed especially for existing wiring.

with the greatest care so as not to split the boards and end up with a poor looking job when the boards are replaced.

If you are in doubt about your ability as a carpenter, it may be best to hire a carpenter to help if a lot of woodwork is involved in installing the electric wiring. Or perhaps you would rather install surface molding as will be described in Chapter 7 for running wiring to light switches.

Outlets installed outdoors or exposed to weather must be protected. Receptacles should have weatherproof covers as shown in Fig. 6-17, and the wiring must be approved for use outdoors. Outlet boxes, connectors, etc. must also be of the type approved for outdoor use.

Type UF cable may be buried directly in the ground for feeding outdoor receptacles. If it is buried less than 18 inches, some means of protection must be provided such as a treated 1- by 8-inch board laid over the cable.

Circuits feeding duplex receptacles on masonry walls may be installed in conduit (Fig. 6-18), surface molding or, in some cases, NM cable.

## CALCULATING RECEPTACLES PER CIRCUIT

Most receptacles used for general use around the home will be fused at no more than 15 amperes and will furnish power to such items as table lamps, television sets, radios, and stereos. Circuits containing such outlets should have no more than eight receptacles on each circuit.

Two small appliance circuits fused at 20 amperes should be provided for the kitchen and dining areas where small

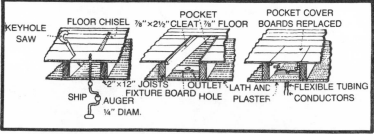

Fig. 6-16. Method of removing floor boards in finished homes.

appliances such as toasters, coffee makers, etc. are likely to be used. Besides these, an additional 20-ampere circuit should be provided for the laundry area for the washer and hand iron. AWG #12 wire is recommended throughout the home.

For other receptacles feeding special appliances such as milking machines, refrigerators, 120-volt air conditioners, etc., check the nameplate rating of each to determine how many receptacles may be installed on each circuit. In general, no more than 1400 watts should be connected to a 15-ampere circuit and no more than 1800 watts to a 20-ampere circuit; 1200 and 1500 watts respectively would be better.

Fig. 6-17. Duplex receptacle with weatherproof cover.

Fig. 6-18. Receptacle circuit installed in conduit on a masonry wall.

Fig. 6-19. Illustration of a ground-fault protector.

Remember on all outside receptacles, the National Electrical Code states:

> For residential occupancies all 120-volt, single-phase 15- and 20-ampere receptacle outlets installed outdoors shall have approved ground-fault circuit protection for personnel....

A ground-fault protector, such as the one shown in Fig. 6-19, installed in your circuit breaker panel board with the outside circuit connected will suffice to meet the code. Such a device is not cheap (about fifty dollars), but what price is a life worth? These devices will immediately de-energize the circuit should a fault occur. This includes a person touching a live wire on the circuit when he is grounded well enough so the shock could kill him.

# Chapter 7
# Installing Light Switches

Many lighting control devices have been developed to make the best use of lighting equipment. These include automatic timing devices for turning lights on and off at desired intervals, dimmers to achieve certain objectives in lighting and of course the common light switch used in nearly every home and farm in the nation. Regardless of the type of lighting the usefulness and the convenience derived from proper switching is well worth the small cost involved.

A wall light switch (Fig. 7-1) is a device used on branch circuits to control lighting and most generally fall into three basic categories:

- Snap-action switches
- Mercury switches
- Quiet switches

The first type consists of a device containing two stationary current-carrying elements, a moving current-carrying element, a handle for the moving element, a spring and an enclosure. When the handle is in the down or off position, as in Fig. 7-2, no current can pass, and the light or lights on the circuit will not burn. When the moving element is closed by moving the handle to the up position, the circuit is complete and the light will burn as shown in Fig. 7-3.

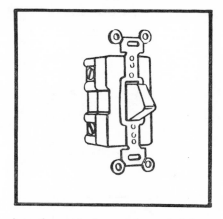

Fig. 7-1. Wall switch.

Mercury switches consist of a sealed capsule containing mercury. A handle is provided to tilt the capsule to allow the mercury to float to one end to bridge two contact points to close the circuit when it is in the on position, and to tilt the mercury away from the contact points when the switch is in the off position. Such switches offer the ultimate in silent operation, but are much higher in price than either the snap-action or quiet switch.

The quiet switch is a compromise between the snap-action switch and the mercury switch. Its operation is considerably quieter than the snap-action switch, yet it is not as expensive as the mercury switch. It is the most commonly used switch for modern lighting practice and are manufac-

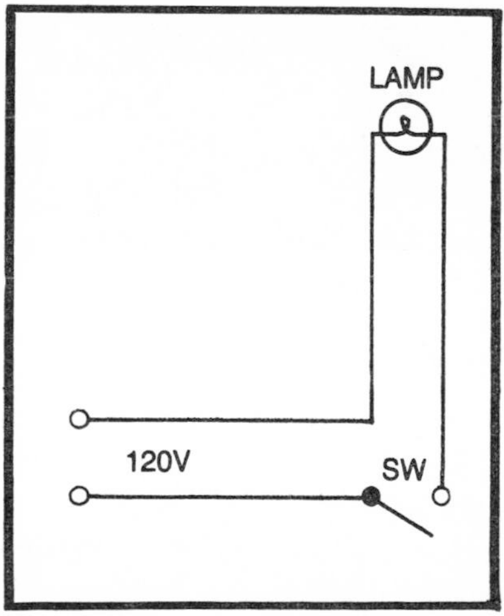

Fig. 7-2. Light circuit with switch in off position.

LAMP

120V    SW

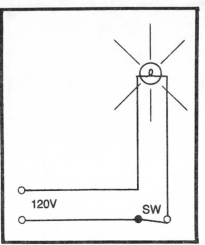

Fig. 7-3. Light circuit with switch in on position.

tured for loads from 10 to 20 amperes in single-pole, two-pole, three-way, four-way, etc. types.

Many other types of switches are available for lighting control. Besides three-way, four-way, and other switches to control lighting from more than one location (see Chapter 8) types such as the door-actuated type are common. The door-actuated type is generally installed in the door jam of a closet to control a light inside the closet. When the door is opened, for example, the light comes on; when the door is closed, the light goes out. The switch on your refrigerator or freezer works in a similar manner.

There was a time when many lights around the home—especially closet, basement and attic lights—were controlled by a pull chain, causing the homeowner to search blindly in the dark for the blasted string. Today, however, there is little excuse not to have wall switches on all lighting fixtures around the home. Even receptacles serving table lamps can be switched at the entrance door to the room.

### INSTALLATION

A wiring diagram of a two-wire 120-volt circuit is shown in Fig. 7-4 feeding a common pull-chain lighting fixture. To install a wall switch, merely locate the switch, pull a piece of two-wire cable from the lighting fixture to the switch and connect according to the wiring diagram in Fig. 7-5. The light

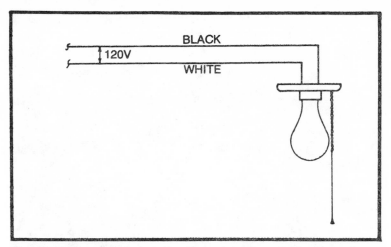

Fig. 7-4. Wiring diagram of a 120-volt circuit feeding a lighting fixture with a pull-chain switch.

fixture can then be controlled as seen in the diagrams in Figs. 7-2 and 7-3.

The basic installation of a wall switch is very similar to installing a duplex receptacle as described in Chapter 6. That is,

1. Locate the outlet box for the wall switch. Make sure that no obstacle is in the wall where the outlet box

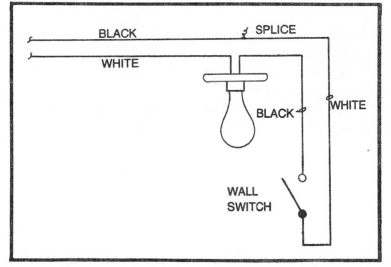

Fig. 7-5. Method of connecting a wall switch to the circuit in Fig. 7-4.

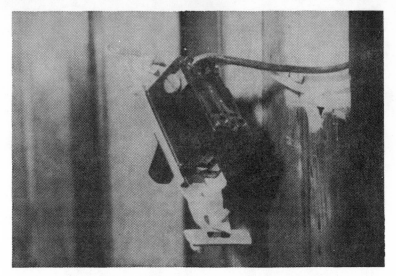

Fig. 7-6. Method of connecting switch-leg wires to the wall switch.

hole is to be cut and that you can route the wires from the switch to the light in an easy manner.

2. Cut out the opening for the switch box. Install and secure the box.
3. Pull the cable from the switch box to the lighting fixture box.
4. Connect the switch leg (wires) to the lighting fixture wires as shown in the diagram in Fig. 7-5.
5. Connect the switch leg to the wall switch as shown in Fig. 7-6.
6. Secure switch to box and install the switch plate. Replace the fixture.

The current should of course be turned off during any portion of this work. The procedure for checking this is described in Chapter 6.

In wiring existing homes, you will encounter the same problems with pulling and fishing wires as was described in Chapter 6. Sometimes, however, it is impractical to run the wires concealed. Therefore, some alternate method must be used. One such method is installing the switch leg and other wires in surface metal raceway, sometimes called surface molding or wire mold. A complete description follows.

## INSTALLING SURFACE METAL MOLDING

When it is impractical to install the wiring in the home concealed surface metal molding is a good compromise. While it is visible proper painting to match the color of the ceiling and walls makes it very inconspicuous.

It is made from sheet metal strips drawn into shape and comes in various shapes and sizes with factory fitting to meet nearly every application found around the home. A few of the fittings available are shown in Fig. 7-7. A complete lists of the fittings can be found by writing the Wiremold Company, whose address is given in Appendix II of this book.

The running of straight lines of surface molding is simple. A length of molding with the coupling slipped in the end is

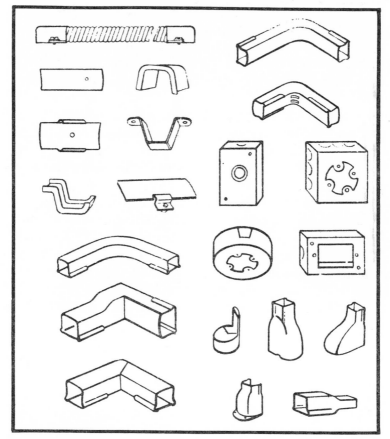

Fig. 7-7. Wire mold fittings available.

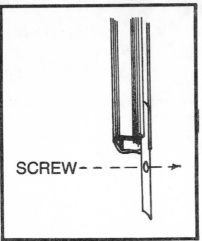

SCREW- - - -

Fig. 7-8. Method of coupling links of wiremold together.

shown in Fig. 7-8. This coupling is slipped out so that the screw hole is exposed, and the coupling is screwed to the surface to which the molding is to be attached (see Fig. 7-9). Then slip another length of molding on the coupling as shown in Fig. 7-10.

Factory fittings are used for corners and turns, or the molding may be bent (somewhat) with a special bender. Matching outlet boxes for surface mounting are also available. Bushings are necessary at such boxes to prevent the sharp edges of the molding from injuring the insulation on the wire.

Clips are used to fasten the molding in place. The clip is secured by a screw, then the molding is slipped into the clip. Wherever extra support of the molding is necessary a strap is slipped over the molding and fastened by screws. When parallel runs of molding are installed, they may be secured in place by means of a multiple strap. The joints in runs of molding are covered by slipping a connection cover over the joints as shown in Fig. 7-11.

Such runs of molding should be grounded the same as any other metal raceway. This is done by use of grounding clips as shown in Fig. 7-12. The current-carrying wires are normally pulled in after the molding is in place.

The installation of surface metal molding requires no special tools unless bending the molding is necessary. The molding is fastened in place with screws, toggle bolts, etc.,

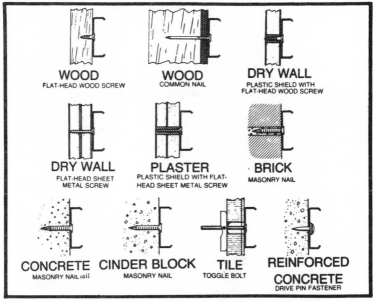

Fig. 7-9. Method for mounting wire mold to different kinds of surfaces.

depending on the materials to which it is fastened. All molding should be run straight and parallel with the room or building lines, that is, baseboards, trims and other room molding. The decor of the room should be considered first, and the molding made as inconspicuous as possible. Figure 7-13 shows an illustration of surface molding, including boxes and fittings, to add a wall switch and related wiring to control a ceiling lighting fixture in an existing home.

It is often desirable to install surface molding not used for wires, in order to complete a pattern set by other surface

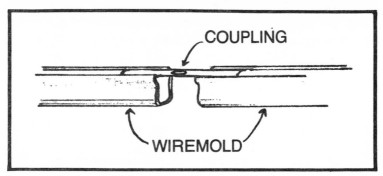

Fig. 7-10. Final step of fitting wiremold links together.

95

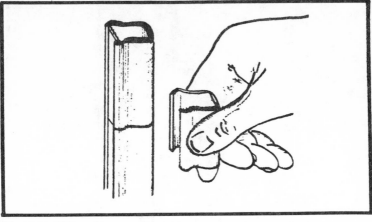

Fig. 7-11. Wire mold connection cover.

molding containing current-carrying wires, or to continue a run to make it appear as a part of the room's decoration.

## INSTALLING A WALL SWITCH IN PLASTERED WALL

The method of installing a duplex receptacle in a wall where one side is open or unfinished was described in Chapter 6, and is a relatively simple operation. The same procedure may also be followed for installing wall switches in a similar area. Only the connections and, of course, the wiring device will be different. However, for a step-by-step method of installing a wall switch, let's take a more complicated problem, such as the installation on an interior first-floor wall of a two-level residence.

A cross section of the house is shown in Fig. 7-14. This section shows both floors of the residence, the interior partitions and their relationship to each other. The existing light fixture is located approximately in the center of the room and is now controlled by a pull-chain switch located in the fixture itself. It is desired to install a new wall switch on the interior wall which is near the entrance door to the room. We can see immediately that the routing of the wires from the light fixture to the switch location is not going to be easy because all of the adjacent surfaces are finished with no immediate access. Surface molding could be run, but it is desired to conceal all wiring.

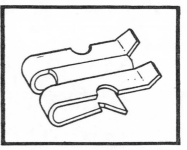

Fig. 7-12. Wire mold grounding clip.

To begin remember that we should plan the installation with the least amount of cutting and patching, and one good procedure is as follows.

Locate the desired spot on the wall for the switch box. Be sure that no stud or other obstacle (water pipes, etc.) are

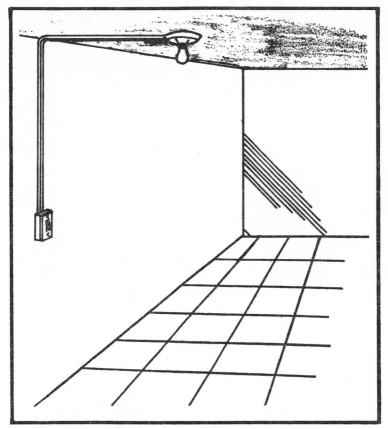

Fig. 7-13. Wire mold used to add a wall switch to a ceiling lighting fixture.

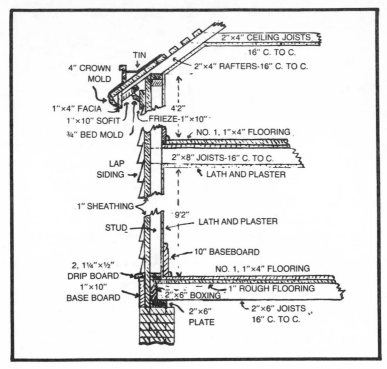

Fig. 7-14. A cross section of a house.

in the partition behind this spot. The box hole should be carefully marked by drawing a pencil around the outlet box, held against the plaster. In locating the exact spot to cut these openings in the plaster it is best to cut a very small hole in the center of the marked area first, using this to locate the cracks between the lath. Then it is possible to shift the mark for the outlet box up or down a little so the lath can be cut properly as shown in Fig. 7-15. Be careful not to cut the hole so large that the switch plate will not cover it neatly.

Once the outlet box hole is cut go to the second floor. Remove the quarter-round molding and baseboard along the floor line above the partition where the outlet box opening was cut. Be extremely careful not to split the wood when removing the trim. Next drill down through the second floor and through the partition plate as shown in Fig. 7-16. Before returning to the first floor guide a fish tape through the drilled hole until the fish tape is near the outlet box opening on the

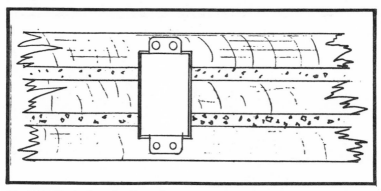

Fig. 7-15. Illustration of wall opening cut for new outlet box. Note the position of the lath.

first floor. It is best if another person assists you with a flashlight on the first floor. When he sees or hears your fish tape near the outlet box opening he can snag it with another fish tape, pulling it out the opening as shown in Fig. 7-17.

Your assistant then hooks his fish tape onto yours and you pull his tape up through the wall partition plate but not

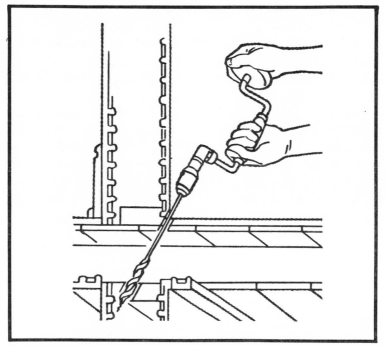

Fig. 7-16. Position for drilling through the second floor to the wall partition below.

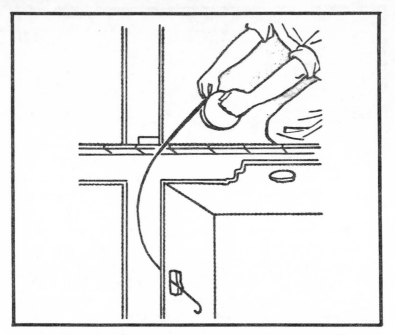

Fig. 7-17. Method of using fish tapes to route wiring in existing partitions.

through the hole in the second floor. Rather, disengage your fish tape from the other and pull yours out, leaving the other tape end between the second floor and the first floor ceiling. Now you can return to the first floor with your fish tape.

Remove the existing lighting fixture and its outlet box. Make certain that the current is shut off first. Then insert your fish tape in the fixture outlet hole to snag the other tape as shown in Fig. 7-18. Then pull it out the fixture outlet opening. You can now easily pull the new cable from the existing lighting fixture to the new wall switch opening.

Secure the cable (two wires with ground wire) to the fish tape by twisting the wires around the fish tape hook, then tape with friction tape. Next with the fish tape, carefully pull the new cable through the void space between the two floors, down through the hole drilled in the partition plate and finally out through the switch box opening. Leave about 18 inches of cable at each end.

Leave the white neutral wire connected to the lighting fixture as it was before you removed it. Remove the black

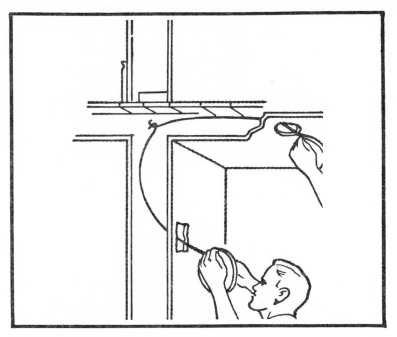

Fig. 7-18. Another method of using fish tapes to help route the electrical wiring.

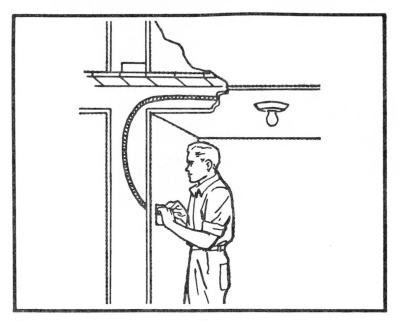

Fig. 7-19. With the switch wires connected to the lighting fixture it's time to start wiring the switch itself.

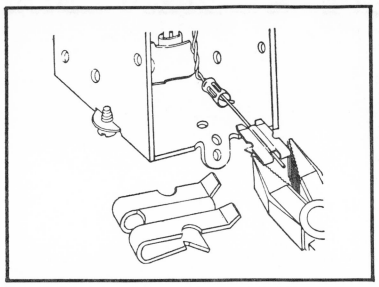

Fig. 7-20. Method of securing ground wire to a metal outlet box.

(hot) wire from the fixture, and splice it to the white wire of the new cable pulled to the new wall switch. Use a wire nut to secure the splice. While you are still at the lighting fixture location connect the black wire of the new cable to the lighting fixture, and replace the fixture to its outlet box (see Fig. 7-19).

Now for the other end of the cable. Strip about 12 inches of sheathing from the cable, then approximately 1½ inches of insulation from the wires inside of the sheathing. Place an outlet box on the ends of the wires, and tighten the cable clamp inside of the box. The box should have plaster ears. Secure the box to the plaster lath with small wood screws (see Fig. 7-15).

Secure the bare ground wire to the metal outlet box with a ground clip as shown in Fig. 7-20. Do this before making an eye in each of the remaining wires and securing each under one of the screw terminals on the switch. The switch is then fastened to the outlet box with the screws provided with the switch itself. Then a wall plate is installed for a finished job.

If you should encounter bridging or a fire stop when fishing the wires inside of a partition, you can cut a hole in the plaster at the bridging, notch the wood, and run the wire through this notch as shown in Fig. 7-21. The hole can then be patched with spackling compound or similar patching material.

## OTHER SWITCH APPLICATIONS

Sometimes it is desirable to install a dimmer control on some of the lights around the home. For example, the dimming of a dining room light can create a mood of candlelight for formal dining. The dimming of recreation room lights provides a good atmosphere for viewing TV or for general conversation. A hall light dimmed at night will serve as a night light without using excessive current.

The dimming of incandescent lamps is best accomplished by reducing the voltage by means of variable transformers, magnetic amplifiers, saturable reactors, or electronic dimmers which are now available for residential use. Most are manufactured in various ratings from 300 to 2000 watts and start at about five dollars for the smaller ratings and jump to over one hundred dollars for the higher ratings.

Most types of residential lighting dimmers can be used to replace a regular wall switch by merely disconnecting the two leads from a single-pole switch, then connecting the two wires to the two screw terminals provided on the dimming device. Therefore, for all practical purposes the installation of most dimmers is exactly like that of a wall switch.

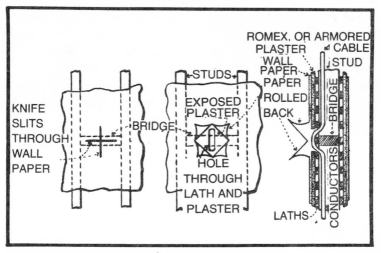

Fig. 7-21. Method of bypassing bridging in a wall partition.

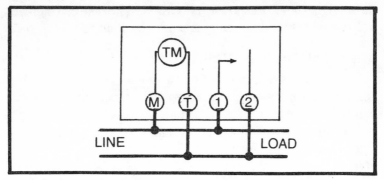
Fig. 7-22. Circuit showing time switch connected to a lighting circuit.

The photocell-controlled lighting circuit is a popular way to add automatic control and convenience to outdoor lighting. They are installed in a manner similar to regular wall switches and complete instructions are included in the packing of each.

Another means of automatic control is the use of electric timers. Timer switches can be set to open and close at predetermined intervals and may be connected to the lighting circuit as shown in Fig.7-22. Several models now available can even compensate for the changing of the number of daylight hours during the seasons and will faithfully perform their function all year long.

Since the development of transistors and other solid-state devices more sophisticated circuit controls have been developed during recent years to perform all kinds of control functions. For example, radio-control circuits can be used to govern lighting just as they are used to control the opening and closing of garage doors. Another application would be to use one of these devices to control the exterior lighting around the home, that is, to turn the outside lights on when a car entered the driveway of your home.

# Chapter 8
# Why Walk in the Dark?

Every room or area in the home having more than one entrance door would have lighting control at each entrance to turn the lights on when entering and to turn them off when leaving. This chapter deals with the various methods of providing this control.

## THREE-WAY SWITCHES

Three-way switches are used to control one or more lamps from two different locations, such as at the top and bottom of stairways, in halls, and other places in the home. Farmers will want to have the convenience of three-way switches from building to building or from the house to any of the other buildings around the farm. For example, the yard or barn lights may be turned on while working around the farm at dusk, then turned off when the occupants return to the house. Or the lights may be turned on at the house and turned off at any of the outer buildings.

Unlike single-pole switches which control a light or group of lights from only one location (containing two wire terminals), three-way switches makes it possible to control lighting from two locations as can be seen in the wiring diagram in Fig. 8-1. By tracing the circuit it may be seen how these three-way switches operate. Two wires are connected to the

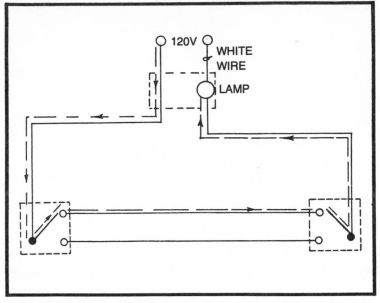

Fig. 8-1. Circuit showing the connections for controlling a lighting fixture from two different locations.

120-volt line. One wire is connected directly to the light fixture (this is the neutral or white wire) while the other black (hot) wire continues on to one of the three-way switches. Now if both handles of the three-way switches are in the up position as in Fig. 8-1 the current will pass through the top *traveler* wire (between the switches) and on through the other switch to the lamp, which will light due to the completed circuit. If either of the handles are turned to the down position, the circuit will be opened, and the lamp will go out. However, the lamp may be turned on again by changing the position of either handle at either switch; that is, if the handle of the left-hand three-way switch is turned to the down position the light will burn; or if the handle on right-hand three-way switch is turned to the up position the light will burn. Thus, the lamp is controlled by switches at two locations.

## FOUR-WAY SWITCHES

When it is desired to control a light or a group of lights from more than two locations one or more four-way switches will have to be added to two three-way switches to accomplish this. In circuits of this type, you will always need two three-way switches—one on each side of the group of four-way switches—and one four-way switch for each additional location. To illustrate, look at the wiring diagram in Fig. 8-2. This shows a lamp controlled from four locations.

In general, the wiring diagram shows a 120-volt, two-wire circuit feeding a lighting outlet. The white or neutral wires connect directly to the terminal on the lighting fixture while the black wire continues to one of the three terminals on one of the three-way switches. This terminal is known as the *point* terminal. Two traveler wires are then connected to the remaining terminals on the three-way switch and run to two of the four terminals on the next four-way switch. Two other traveler wires are connected to the other two terminals on this four-way switch and run to the other four-way switch. The two remaining traveler wires connect to the other three-way switch. Then one wire from the point terminal connects to the other side of the lighting fixtures. The actuation of any one of these four switches will turn the light on or off.

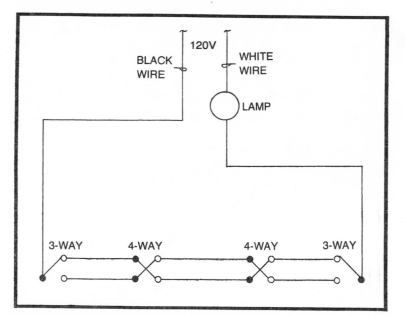

Fig. 8-2. Wiring diagram of lighting fixture controlled from four different locations.

## THREE-WAY SWITCHING SYSTEM INSTALLATION

Let's assume that you desire to install a lighting circuit, controlled by two three-way switches, from your house to an outbuilding such as a garage or a barn. You have decided to run the wiring on poles overhead, and the wiring will consist of several single-conductor wires, not a multiconductor cable. An elevation drawing of the project is shown in Fig. 8-3.

First, the electric load in the barn must be examined to determine the proper wire size to be used for the run. We find that seven 100-watt lamps will be installed in the barn along with one duplex receptacle that will be used periodically. It is estimated that the maximum load that will be used on this outlet at any one time will be 700 watts (found on the nameplate of the largest tool that will be used. Since the line voltage will be 120 volts, the current, in amperes, may be found by the formula,

$$\text{Current} = \frac{\text{total load in watts}}{\text{voltage}}$$

Therefore, the current will be 1400/120 = 11.66 amperes. It would be natural for us to then choose a wire size from tables in the National Electrical Code that are rated for this current, but there are other factors that must be taken into consideration. One factor is the distance of 200 feet. At the given load and this distance a wire size must be selected that will keep the normal voltage (120 volts) within two percent of its original rating, that is, 120 × 0.02 = 2.40. Therefore, 120–2.40 = 117.6 volts.

There are formulas used by electrical designers and engineers to determine the voltage drop in circuits, but for the average homeowner, voltage-drop tables such as the one in Table 8-1 is easier and quicker to use.

We previously calculated the load for our circuit in question to be 11.66 or 12 amperes. Referring to the left-hand column in Table 8-1 we find *12* amperes. Then continue to the right until we come to the column under *200*—the length of feet in the run. We can see that the table calls for AWG #6 to carry a load of 12 amperes based on a two percent voltage drop. Therefore, we will use AWG #6 copper wire that is suitable for outdoor use.

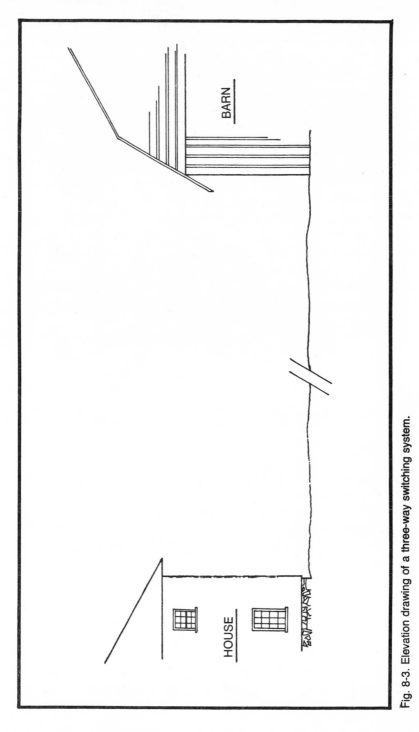

Fig. 8-3. Elevation drawing of a three-way switching system.

The next problem is that of supporting the cable through the air from the house to the outbuilding. The distance is too far to attach one end of the wires to the house and the others to the out building. There must be some support in between.

A rule of thumb for spacing poles is the poles should be spaced approximately 125 feet apart for straight runs. Since the distance from the house to the barn is 200 feet in our example, one pole set halfway between the two buildings would suffice.

The size of the pole cannot be definitely specified here because of the varying conditions. However, in general, the pole should be not less than 24 feet above the ground. It should be high enough to allow loaded trucks and other vehicles to pass well under the electric lines. The lowest wire on the pole must be at least 18 feet above the ground if the wires cross a driveway. If the wires do not pass over anywhere that traffic is anticipated then the lowest wire on the pole may be only 12 feet above the ground.

A 30-foot pole should be set in the ground approximately 6 feet deep, leaving 24 feet above ground. Shorter poles can be less. For example, 3 feet is fine for a pole 18 feet in overall length. This leaves 15 feet of pole above ground and would be satisfactory for running the wires from the house to the barn if no traffic passed under the lines.

When digging the hole for the pole it should be large enough to take the pole and allow for a tamping bar to be worked around pole to tamp the loose soil.

Figure 8-4 shows the basic components and wiring of the system. A two-wire circuit is fused at the fuse cabinet and run to an outlet box containing the three-way switch in the house. The hot wire of this circuit is connected to the point terminal on the three-way switch while the neutral wire bypasses the switch. Two traveler wires and the neutral are run in rigid conduit from the switch box up the side of the house high enough to obtain the desired height above ground. A weatherhead is provided on top of the conduit to prevent water from entering the pipe.

A bracket with three porcelain insulators are attached to the conduit to support the wire span from the house to the pole, and the conduit is securely strapped to the house.

Table 8-1. AWG/MCM Based on 2% Voltage Drop

| Length (ft) | 30 | 40 | 50 | 60 | 70 | 80 | 90 | 100 | 125 | 150 | 175 | 200 | 225 | 250 | 275 | 300 | 350 | 400 |
|---|---|---|---|---|---|---|---|---|---|---|---|---|---|---|---|---|---|---|
| | **Amper Load** | | | | | | | | AWG/MCM* | | | | | | | | | |
| 5 | 14 | 14 | 14 | 14 | 14 | 14 | 12 | 12 | 12 | 10 | 10 | 10 | 8 | 8 | 8 | 8 | 6 | 6 |
| 6 | 14 | 14 | 14 | 14 | 14 | 12 | 12 | 12 | 10 | 10 | 10 | 8 | 8 | 8 | 8 | 6 | 6 | 6 |
| 7 | 14 | 14 | 14 | 14 | 12 | 12 | 12 | 12 | 10 | 10 | 8 | 8 | 8 | 8 | 6 | 6 | 6 | 6 |
| 8 | 14 | 14 | 12 | 12 | 12 | 12 | 12 | 10 | 10 | 8 | 8 | 8 | 8 | 6 | 6 | 6 | 6 | 4 |
| 9 | 14 | 14 | 12 | 12 | 12 | 10 | 10 | 10 | 8 | 8 | 8 | 8 | 6 | 6 | 6 | 4 | 4 | 4 |
| 10 | 14 | 14 | 12 | 12 | 12 | 10 | 10 | 10 | 8 | 8 | 6 | 6 | 6 | 6 | 6 | 4 | 4 | 4 |
| 12 | 14 | 12 | 12 | 10 | 10 | 10 | 10 | 8 | 8 | 6 | 6 | 6 | 4 | 4 | 4 | 4 | 4 | 2 |
| 14 | 14 | 12 | 10 | 10 | 10 | 8 | 8 | 8 | 6 | 6 | 6 | 6 | 4 | 4 | 4 | 4 | 2 | 2 |
| 16 | 12 | 12 | 10 | 10 | 8 | 8 | 8 | 6 | 6 | 6 | 4 | 4 | 4 | 4 | 4 | 2 | 2 | 2 |
| 18 | 12 | 12 | 10 | 10 | 8 | 8 | 8 | 6 | 6 | 4 | 4 | 4 | 4 | 2 | 2 | 2 | 2 | 1 |
| 20 | 12 | 10 | 10 | 8 | 8 | 6 | 6 | 6 | 4 | 4 | 4 | 2 | 2 | 2 | 2 | 2 | 1 | 0 |
| 25 | 10 | 10 | 8 | 8 | 6 | 6 | 6 | 4 | 4 | 2 | 2 | 2 | 2 | 1 | 1 | 1 | 0 | 00 |
| 30 | 10 | 8 | 8 | 8 | 6 | 6 | 4 | 4 | 2 | 2 | 2 | 1 | 1 | 1 | 0 | 0 | 00 | 00 |
| 35 | 8 | 8 | 6 | 6 | 4 | 4 | 4 | 4 | 2 | 2 | 1 | 1 | 0 | 0 | 0 | 0 | 00 | 000 |
| 40 | 8 | 6 | 6 | 6 | 4 | 4 | 4 | 2 | 2 | 1 | 1 | 0 | 0 | 00 | 00 | 00 | 000 | 000 |
| 45 | 6 | 6 | 4 | 4 | 4 | 2 | 2 | 2 | 1 | 1 | 0 | 0 | 00 | 00 | 00 | 000 | 000 | 0000 |
| 50 | 6 | 4 | 4 | 4 | 2 | 2 | 2 | 1 | 0 | 0 | 00 | 00 | 000 | 000 | 000 | 0000 | 0000 | 250 |
| 60 | 4 | 4 | 2 | 2 | 2 | 1 | 1 | 0 | 00 | 00 | 000 | 000 | 000 | 0000 | 0000 | 0000 | 250 | 300 |
| 70 | 4 | 2 | 2 | 1 | 1 | 0 | 0 | 00 | 000 | 0000 | 000 | 0000 | 0000 | 250 | 250 | 250 | 300 | 350 |
| 80 | 2 | 1 | 1 | 0 | 0 | 00 | 00 | 000 | 0000 | 250 | 300 | 250 | 300 | 300 | 350 | 300 | 300 | 400 |
| 90 | | | 0 | | 00 | 000 | 000 | 0000 | 250 | 300 | 350 | 300 | 350 | 400 | 400 | 350 | 500 | 500 |
| 100 | | | | | | | | | | | | 400 | 500 | 500 | 500 | 500 | 500 | |
| 130 | | | | | | | | | | | | | | | | | | |
| 160 | | | | | | | | | | | | | | | | | | |
| 200 | | | | | | | | | | | | | | | | | | |

Another bracket is attached to the pole as well as to the barn to support the wire span.

At the barn, a three-wire cable is tapped to the overhead span and run to another outlet box containing another three-way switch. Notice that the neutral does not connect, in any way, to the switch. Rather, it continues on with the wire coming from the point terminal on this switch to feed the light and duplex receptacle.

With this wiring arrangement the lights have to be burning if it is desired to use the receptacle, but since the receptacle will be used infrequently, this should cause no problem. If it is desired to keep the receptacle hot all of the time and control the lights independently, then another wire would have to be added to the system. This wire could be connected to the point terminal on the three-way switch in the house, then run with the other wires to the barn. At the barn, this extra wire would feed the receptacle directly without running through the three-way switch. The neutral wire could be tapped from the neutral feeding the lights in the barn since it bypasses the three-way switch also.

## LOW-VOLTAGE REMOTE-CONTROLLED SWITCHING

In many homes and on farms where lighting must be controlled from several points, or where flexibility is desirable in the lighting system, low-voltage remote-controlled relay systems have been applied. These systems use standard 120-volt circuits to feed the lights, but the relay switches are controlled by low-voltage components, operated from a transformer. Because the control wiring itself does not carry the lighting load directly small lightweight doorbell-type cable can be used. It can be run wherever and however convenient—placed behind molding, stapled to woodwork, buried in shallow plaster channels or installed in holes bored in wall studs or ceiling and floor joists. Outlet boxes are not even required for terminations when using this system.

The basic circuit of a remote-control switching system is shown in Fig. 8-5. A regular two-wire 120-volt circuit is used to feed the lighting fixture outlet box. The white wire or neutral is connected directly to the fixture. The black or hot

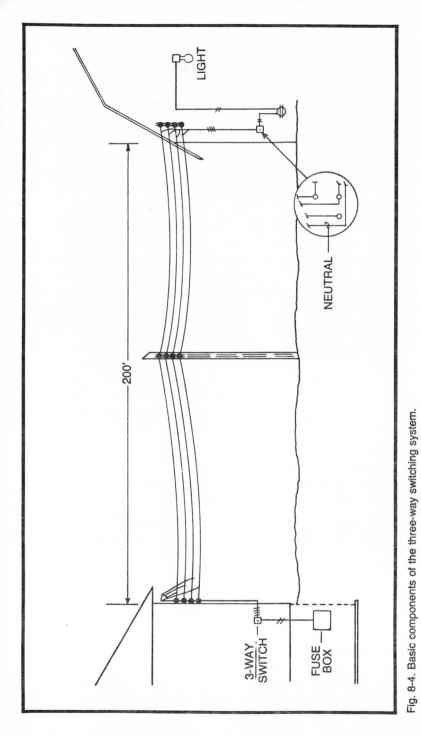

Fig. 8-4. Basic components of the three-way switching system.

LIGHT

NEUTRAL

200'

3-WAY
SWITCH

FUSE
BOX

wire bypasses the fixture and connects to a small relay attached to the outlet box in a way similar that it would be connected to a regular single-pole wall switch. Another wire is connected to the other side of the relay, then connected to the light fixture. Therefore, when the relay contacts are closed (to complete the circuit) the light will burn; when opened the light will not burn. However, a means of opening and closing the relay automatically is necessary for a proper installation.

A small 120/24-volt transformer is connected to the 120-volt lines with two of its primary leads. Then the blue and white secondary leads (on the opposite side of the transformer) will produce 24 volts. The hot or blue wire is connected to one side of the magnetic coil in the relay, and the white or common wire from the transformer is connected to the middle or common terminal on a low-voltage wall switch. A red wire is connected between the on terminal of the switch and the on terminal of the magnetic coil. A black wire is then connected between the off position of the switch and the off position of the coil. This completes the wiring.

Figure 8-5 shows the contacts in an open or off position, but by pressing the low-voltage switch so that the common (white) wire comes into contact with the red wire, thus completing the circuit through the on coil, the iron core of the relay will be drawn under the on coil and cause the contacts to close. In doing so, the 120-volt circuit is also completed and the light will burn.

When it is desired to turn the light off, merely press the low-voltage switch to the off position. This will complete the circuit between the white and black wire causing the iron core to move back to the off position and opening the 120-volt circuit.

## APPLICATION OF LOW-VOLTAGE SWITCHING

Let's assume that it is desired to control a large hall lighting fixture from each of the six doors entering the hallway. If conventional methods were used, this would take two three-way switches, and four four-way switches, plus many feet of two- and three-wire AWG #12 copper cable. Since the structure is existing, this means cutting and patching the walls

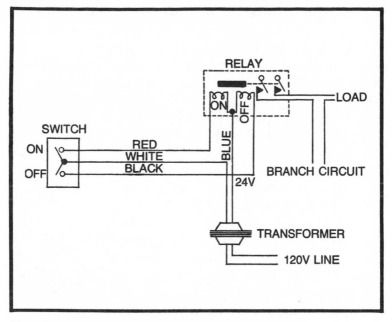

Fig. 8-5. Basic circuit of a remote-controlled switching system.

and ceiling to run the large cable through. However, by using the low-voltage bell wire, most of the wiring can be run behind molding or door trim which would eliminate most of the cutting and patching.

This project (Fig. 8-6) would require one low-voltage relay, one 120/24-volt transformer, six low-voltage switches, bell wire and miscellaneous hardware. We would begin the project as follows.

Pull the fuse or trip the circuit breaker feeding the light fixture to completely deenergize the circuit. Then remove the light fixture and loosen the box so that you can remove one of the knockouts from the outlet box in order to insert and secure the low-voltage relay. The line side of the relay is made to fit a ⅜-inch knockout. A locknut screwed on from the inside of the outlet box will hold the relay secure to the box.

If there is enough room in the box the small transformer may also be inserted in a knockout on the opposite side of the outlet box from the relay. Again, most transformers of this type have their line-side wires arranged through a threaded hub for inserting it into a regular ⅜-inch outlet box knockout.

Once both the relay and transformer are secure connect the black (hot) wire feeding the outlet box from the fuse cabinet or panelboard to one leg of the relay, and also tie in one leg of the transformer. Twist all three leads around each other in a rattail splice. Then use about a 76B wirenut to secure the splice.

Next splice the other lead from the transformer line-side wire to the white or neutral conductor of the 120-volt circuit feeding the outlet box. However, do not put a wire nut on this splice as the fixture will also have to be spliced to this wire.

While you're up on the ladder you may as well connect the blue lead on the low-voltage side of the transformer to the proper terminal on the relay. This connection may be made with the rattail splice and secured with a 72B wirenut.

Now by hook or crook, run a piece of three-wire bell cable from the relay/transformer location to the first low-voltage wall switch location. The cable may be fished down through the partitions by methods described in Chapters 6 and 7, installed in surface metal molding or run behind or tacked on wooden molding. Leave about 1 foot of cable at each end in order to make splices.

Go back to the outlet box where the relay and trans-former are located. Connect the red and black wires of the cable to the relay in their appropriate places. Next connect the white wire in the cable to the remaining white wire on the low-voltage side of the transformer. Replace and secure the outlet box after you are certain that all of the connections are correct and securely made. Now you're ready to replace the light fixture. First connect the white lead from the fixture to the white wire in the 120-volt circuit feeding the outlet box and to which one of the line-side leads from the transformer is connected. Secure this splice with a wire nut.

Now with a short piece of wire the same size as was used in the 120-volt circuit connect one end to the remaining terminal on the line side of the relay and the other end to the black wire on the light fixture. After these joints are made secure replace the light fixture. Make certain that this is secured in a manner opposite from the way you removed it. This ends the work at the outlet box and fixture, but now back to the low-voltage switches.

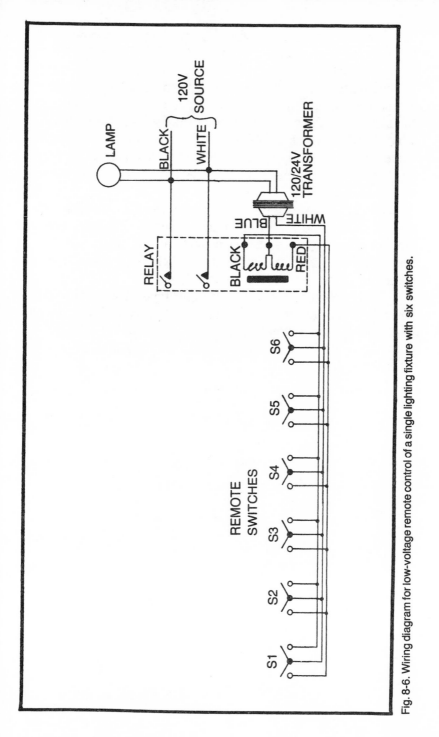

Fig. 8-6. Wiring diagram for low-voltage remote control of a single lighting fixture with six switches.

You now have a three-wire cable pulled to the location of the nearest low-voltage switch to the light fixture outlet. However, before connecting the switch to these wires run another piece of three-wire bell cable from this location to the location of the next nearest low-voltage switch. Now splice the six ends of the cable together at the first switch, that is, white to white, black to black, and red to red. This again is made with the rattail splice. Now insert each pair of wires under the proper screw terminal on the switch and secure the switch to the wall.

Continue the procedure given in the preceding paragraph until all six of the low-voltage switches are wired. In fact, any number of low-voltage switches could be added or deleted to this three-wire cable to control the one light fixture from any point one of the switches are located.

The method just given is for one type of low-voltage control system produced by one manufacturer. While the wiring may be slightly different on systems manufactured by other firms the basic components and wiring will be essentially the same.

# Chapter 9
# Ready to Tackle 240-Volt Circuits?

There are many electrical appliances and equipment around the home and farm that require 240 volts to operate. Some of these include electric ranges, clothes dryers, water heaters, air conditioners, electric heating equipment, electric motors driving grain conveyors, milking machines, electric welders and similar pieces of equipment. Many homeowners, however, are somewhat reluctant to tackle any wiring job requiring more than 120 volts. This is probably due to fear of being shocked by the higher voltage, or else many think that the higher the voltage, the more complicated the job.

Actually there is no more danger working on a dead 240-volt line than it is on a dead 120-volt line. Both should be handled with care and alive both can certainly harm or even kill a person under the right conditions. On the other hand, if you have taken the necessary precautions before working on any electrical line, there is little danger involved in your working on the lines. Just make certain that the fuses are pulled or the circuit breaker tripped before working on or handling any electrical wiring. Then double-check all open wiring with your testing instrument or test lamp before touching any electrical component which could be hot.

The complexity of 240-volt equipment is no more than any other voltage. In fact, most connections to 240-volt equip-

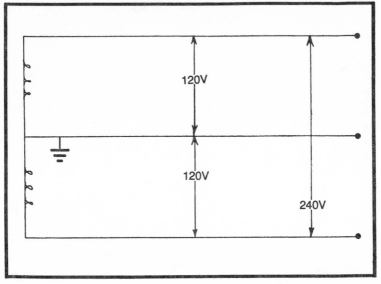

Fig. 9-1. Wiring diagram of a 120/240-volt circuit showing the voltages between phases.

ment involves only the connection of two wires, which is much the same as connecting a typical lighting fixture or a duplex receptacle. On a 120-volt circuit a voltmeter reading taken between the hot wire to any ground will be 120 volts. The neutral wire to ground should read 0 volts. A reading between the two wires—hot and neutral—will read 120 volts. On a 240-volt circuit, a reading between either of the two wires to ground will be 120 volts, while a reading taken between the two live wires will be 240 volts. Figure 9-1 further illustrates this fact.

## LAYOUT OF 240-VOLT CIRCUITS

With certain restrictions, the wiring methods used for conventional branch circuits may be utilized for 240-volt circuits also. These include, nonmetallic cable, armored cable, conduit, etc. The wires, of course, must be of sufficient size to meet the requirements of the National Electrical Code, and are calculated on the basis of the power equipment or appliance load requirements.

For example, in order to determine the size of the wires for an electric range, proceed as follows:

1. Find the nameplate rating of the electric range. This will normally be rated in kilowatts. Assume 12 kilowatts (kW).
2. Refer to the Table 9-1. The *Max Demand* column applies to ranges rated 12 kilowatts and less.
3. Under *Appliances* column locate the appropriate number of appliances (assume only one in our case). Find the maximum demand given for it. The *Max Demand* column states that the circuit should be sized for 8 kilowatts and not for the maximum of 12 kilowatts.

The reason for this last statement in step 3 is that it is unlikely that all of the burners will be operating on high at any one time, especially with the oven on and the other electrically operated devices such as lights, small appliance outlet, etc. Therefore, the table is allowing for a certain amount of diversity in sizing the circuit feeding the electric range. In this case, the wire may be sized for an estimated continuous load of only 8 kilowatts rather than the nameplate rating of 12 kilowatts. Calculate the required size of the circuit wiring as follows:

$$\text{Ampere rating} = \frac{\text{demand load in watts}}{\text{rated voltage}}$$

thus,

$$\frac{8000 \text{ watts (8 kW)}}{240 \text{ volts}} = 33.33 \text{ amperes}$$

Table 9-1. Power Demand of Household Electric Ranges

| Appliances | Max Demand (kW) |
|:----------:|:---------------:|
| 1 | 8 |
| 2 | 11 |
| 3 | 14 |
| 4 | 17 |
| 5 | 20 |

In checking the current carrying capacity of copper wires in Table II-1 in Appendix II, we find that no wire size is listed for exactly 33.33 amperes. Therefore, we will use the next higher size which is rated at 40 amperes and is AWG #8.

When sizing wires for most of the other 240-volt appliances used in the home or around the farm they usually must be sized for their full capacity because there is no diversity or demand factor for most of these appliances. In fact, the wire size for most other 240-volt appliances or equipment should be sized at 125 percent of their nameplate rating. For example, if we were sizing the feeder wires for a 4000-watt electric heater, we would find the load in amperes by

$$\frac{4000 \text{ (watts)}}{240 \text{ (volts)}} \times 1.25 = 20.82 \text{ amperes}$$

Since the table in Appendix II gives no wire size for exactly 20.82 amperes the next higher rating would be for 30 amperes, which is AWG #10 copper wire. When sizing wires and the calculated ampere rating does not exactly match one in the table, never go smaller (regardless how close the figure may be), but always use the next *higher* rating.

Another point to consider is the nameplate rating of electric water heaters. There are many types and sizes or residential water heaters from a single 1000-watt element to two 4500-watt elements or higher contained in the tank. Let's assume that a typical residential water heater's nameplate indicates two heating elements of 4500 watts each. Our first consideration would be to size the feeder wires for 9000 watts. However, in most cases, if you examine the wiring diagram of this type of water heater, you will find that only one of the elements will be operating at any one time. The thermostats and relays are connected so that the lower 4500-watt heating element becomes energized first when the thermostat calls for heat, and at the same time, opens a contact to prevent the upper heating element from operating. When the lower-element thermostat is satisfied, the lower contact opens, and at the same time closes the upper thermostat so that it may be energized when its thermostat calls for heat to

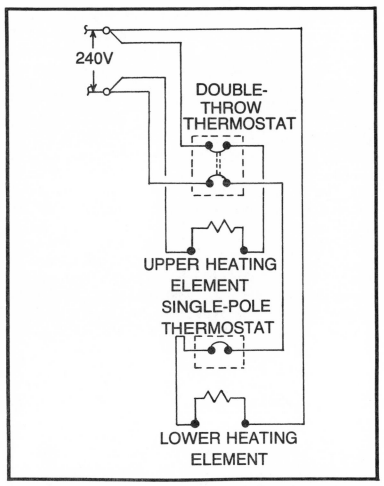

Fig. 9-2. Wiring diagram of a water heater electrical control circuit.

maintain the water temperature. The wiring diagram in Fig. 9-2 more clearly shows the operation of the two elements.

With this information we can now size the wires for the water heater by the formula:

$$\frac{4500 \text{ (watts)}}{240 \text{ (volts)}} = 18.75 \times 1.25 = 23.43 \text{ amperes}$$

Since no wire size is made to carry exactly 23.43 amperes the next highest wire size will have to be used, that is, AWG #10 which is rated for 30 amperes.

Most of the remaining 240-volt appliances should have their wire sized according to their nameplate rating times 125 percent. Still, there is another factor that must be considered when sizing wire for any electrical system—voltage drop. When the voltage drop, the load in amperes and the length of the circuit are known the size of wire to be used can be found by using tables that indicate the proper size of wire to be selected at a given voltage for a calculated load, at a given distance and the allowed voltage drop in percent.

One voltage drop listing was found in Table 8-1 of Chapter 8. This table is based on an allowable voltage drop of two percent and the use of copper wires. If aluminum wires are used instead of copper multiply the current carrying capacity by a factor of 0.84.

Another voltage drop listing for 240-volt circuits is shown in Table 9-2. This table is based on a three percent allowable voltage drop in the circuits and uses copper wires for the examples. The current carrying capacities in this table should also be multiplied by a factor of 0.84 should aluminum wires be used instead of copper.

The importance of voltage drop in circuits should be fully understood by the homeowner and farmer. Take, for example, circuits feeding electric poultry incubators and brooders where the life of the chicks are at stake as well as the farmer's income. If a brooder was rated at 1500 watts at normal voltage, a three percent drop in voltage will cause approximately six percent loss in heating. This means if only 232 volts reaches the brooder only 1410 watts of heat will be produced instead of the rated 1500 watts. While a voltage drop of only three percent will normally not cause any real problems, we have seen some rural electrical systems having more than a seventeen percent voltage drop. With this much voltage drop the 1500-watt brooder would only be putting out approximately 990 watts—enough loss to cause harm to the chicks under some conditions.

There are several ways to insure that the voltage drop will not exceed 3 percent. First, check the voltage at your main electric switch to ascertain that the power company is providing you with voltage between 230 to 240 volts. If not, ask them to adjust their transformer taps so that the voltage

## Table 9-2. AWG/MCM Based on 3% Voltage Drop

| Length (ft) | \|← Amperes | | | | | | | | | | | | | | →\| |
|---|---|---|---|---|---|---|---|---|---|---|---|---|---|---|---|
| | 40 | 50 | 60 | 70 | 80 | 90 | 100 | 125 | 150 | 175 | 200 | 250 | 300 | 350 | 400 |
| | AWG/MCM* | | | | | | | | | | | | | | |
| 5 | 14 | 14 | 14 | 14 | 14 | 14 | 14 | 14 | 14 | 14 | 14 | 14 | 12 | 12 | 12 |
| 6 | 14 | 14 | 14 | 14 | 14 | 14 | 14 | 14 | 14 | 14 | 14 | 12 | 12 | 12 | 10 |
| 7 | 14 | 14 | 14 | 14 | 14 | 14 | 14 | 14 | 14 | 14 | 12 | 12 | 12 | 10 | 10 |
| 8 | 14 | 14 | 14 | 14 | 14 | 14 | 14 | 14 | 12 | 12 | 12 | 12 | 10 | 10 | 10 |
| 10 | 14 | 14 | 14 | 14 | 14 | 14 | 14 | 14 | 12 | 12 | 12 | 10 | 10 | 8 | 8 |
| 12 | 12 | 14 | 14 | 14 | 14 | 14 | 14 | 12 | 12 | 12 | 10 | 10 | 8 | 8 | 8 |
| 16 | 12 | 12 | 12 | 12 | 12 | 12 | 12 | 12 | 10 | 10 | 10 | 8 | 8 | 8 | 6 |
| 20 | 10 | 12 | 12 | 12 | 12 | 12 | 12 | 10 | 10 | 8 | 8 | 8 | 6 | 6 | 6 |
| 25 | 10 | 10 | 10 | 10 | 10 | 10 | 10 | 8 | 8 | 8 | 8 | 6 | 6 | 6 | 6 |
| 30 | 10 | 10 | 10 | 10 | 10 | 8 | 8 | 8 | 8 | 6 | 6 | 6 | 4 | 4 | 4 |
| 40 | 10 | 10 | 10 | 10 | 10 | 8 | 8 | 8 | 6 | 6 | 6 | 4 | 4 | 4 | 4 |
| 50 | 8 | 10 | 10 | 8 | 8 | 8 | 6 | 6 | 6 | 4 | 4 | 4 | 2 | 2 | 2 |
| 60 | 6 | 8 | 8 | 8 | 6 | 6 | 6 | 6 | 4 | 4 | 4 | 2 | 2 | 2 | 2 |
| 80 | 6 | 6 | 6 | 6 | 6 | 4 | 6 | 4 | 4 | 2 | 2 | 2 | 1 | 1 | 1 |
| 100 | 4 | 6 | 6 | 6 | 4 | 4 | 4 | 4 | 2 | 2 | 2 | 1 | 0 | 0 | 0 |
| 130 | 2 | 4 | 4 | 4 | 2 | 2 | 2 | 2 | 2 | 1 | 1 | 0 | 00 | 00 | 00 |
| 160 | 1 | 2 | 2 | 2 | 1 | 1 | 1 | 1 | 1 | 0 | 0 | 00 | 000 | 000 | 000 |
| 200 | 00 | 1 | 1 | 1 | 00 | 00 | 00 | 00 | 0 | 00 | 00 | 000 | 0000 | 0000 | 0000 |
| 250 | 000 | 00 | 00 | 00 | 000 | 000 | 000 | 000 | 00 | 000 | 000 | 0000 | 250 | 250 | 250 |
| 300 | 0000 | 000 | 000 | 000 | 0000 | 0000 | 0000 | 0000 | 000 | 0000 | 0000 | 250 | 300 | 350 | 300 |
| 350 | 250 | 0000 | 0000 | 0000 | 250 | 250 | 250 | 250 | 0000 | 250 | 250 | 300 | 350 | 400 | 400 |
| 400 | | 250 | 250 | 250 | | | | | 250 | | 250 | 300 | 400 | 500 | 500 |

will be up to normal. Next, make certain that your branch circuits are not overloaded. If so, split the load by adding additional circuits. Finally, make certain that the various wire sizes in your electrical system are large enough for the load and the distance to the load. If not, take corrective measures.

From the previous paragraphs, it should now be obvious that the objective in any electrical circuit is to select and install a wire size that will carry the required load without be over-fused, without becoming overheated and with not more than 3 percent voltage drop. The following examples will illustrate some of the common problems in selecting and installing 240-volt circuits in the home and around the farm.

## PRACTICAL APPLICATIONS

Assume that an electric milk cooler is located 125 feet from an electric switch with a voltage of 230 volts. If the nameplate on the cooler gives a full-load current of 27.5 amperes at 230 volts, what size wires would be required to feed the machine and not exceed the three percent maximum voltage drop?

Remember that this type of circuit should be sized at 125 percent of its name plate rating. So, 27.5 amperes × 1.25 = 34.4 amperes. Table II-1 in Appendix II of this book tells us that AWG #8 would be sufficient to carry the load. But what about the voltage drop?

Referring to Table 9-2 look down the left-hand column under *Amperes* until we come to *40*; this is the closest the table comes to 34.4 amperes without going under. Now continue looking across and to the right until we come under the *125* feet column. Since the wire size given is AWG #8 our original selection was correct. However, if the run was 150 feet we would have to jump to AWG #6 to keep the voltage drop within the desired three percent level.

You have just decided to replace your old gas range with a new electric range which will be located approximately 40 feet away from your electric panel. The nameplate rating of your new range is 12 kilowatts. How would you go about installing the 240-volt electric circuit for this range?

If you remember we showed how to determine the wire size for a similar range earlier in this chapter. It would require a minimum of AWG #8 copper wire. Now let's look at Table 9-2 to see if this size will suffice to keep the voltage drop within three percent. In this case, we will use the demand load of 8000 watts instead of the nameplate rating of 12,000 watts.

Since the voltage drop in Table 9-2 is based on loads in amperes rather than kilowatts we must first convert the kilowatts to amperes by dividing 8000 watts by 240 volts, which gives us 33.3 amperes. Then referring to Table 9-2 we find that AWG #8 wires are plenty large enough for this circuit.

As mentioned previously there are several different wiring methods that could be used for this circuit, but let's assume that most residential wiring is done in NM cable and we will also use NM cable for the range circuit. But what type shall we get? In most 240-volt circuits, only two wires are needed to supply the current, but on some 240-volt appliances (this range being one of them) 120 volts are also needed to light lights, run 120-volt clocks, etc. Therefore, this circuit will require two hot wires, a neutral, and a grounding conductor. So we will order three-wire AWG #8 NM cable with a ground wire. The length should be sufficient to allow approximately 2 to 3 feet on each end.

Next run the cable from the panelboard to the location of the electric range by methods described elsewhere in this book, that is, the shortest and easiest way possible with the least amount of cutting and patching room finishes within the house. The cable should be secured a maximum of every 4½ feet by the proper size of cable straps or by pulling the cable through holes bored in studs or joists.

Once the cable is pulled in and secured begin on the end of the cable at the range, and make that connection first. If the range is of the free-standing type you will probably want to install a range receptacle, then use a range cord to plug the range into the receptacle. In this way the range may be unplugged and moved out of its area for periodic cleaning the wall and floor where the range is standing or perhaps to clean the range itself. If the range is one of the built-in types a direct connection to the junction box provided on the range will suffice.

A three-pole, four-wire surface-mounted range receptacle is shown in Fig. 9-3. This type of receptacle has a built-in cable clamp assembly mounted at the bottom knockout and comes complete with instruction for installing. To connect the cable to the receptacle strip about 6 inches of sheathing from the end of the cable with a small knife. Be careful not to cut deep enough to cut through the insulation on the wires inside of the sheathing. Now strip about 1 inch of insulation from each of the wires inside of the sheathing as described in Chapter 3.

Insert all wires through the bottom knockout on the receptacle base until the cable sheathing is approximately a quarter-inch pass the cable clamp provided. Tighten the cable clamp and secure the receptacle base to the wall or floor behind the range.

The terminals should be marked on the receptacle or in the instructions that come with it. But, in general, the white (neutral wire) should be connected to the top center terminal and is secured with a set screw provided with the receptacle. The black wire is connected to one of the side terminal slots, and the red wire is connected to the opposite side terminal slot—both with set screws provided. The grounding wire is then connected to the bottom terminal, whose slot is shaped differently from the others. When all terminals are tightly secured install the cover with the one screw provided. You may now plug in the range cord and move the range in place.

Now for the other end of the cable, the one at the fuse box or panelboard. Here you will need about 2 feet of sheathing removed from the cable to allow the inside wires to be connected to their proper terminal inside of the cabinet. Again the sheathing is removed; however, before doing anything on this end turn the main switch off so that the bus bars and terminals inside of the cabinet will be completely dead. Have a flashlight handy in case you need it when the power is shut off. If the overcurrent protection is of the circuit breaker type, make sure the main breaker is in the off position. Then with a test lamp or testing instrument check all terminals by holding one lead to the neutral block. Next move the other lead from terminal to terminal until you are certain that the panel is dead.

Fig. 9-3. Illustration of a 50-ampere range receptacle.

Remember, however, that the large service wires feeding the main breaker in the panel will still be hot unless the meter is pulled. This should not make any difference as you won't have to get near them. Just be careful that you don't. As a further precaution, place a piece of plywood on the floor for you to stand on. If you should happen to brush your hand against one of the live wires, and if you're not touching any grounded object, you won't get any shock.

With the cover of the panelboard removed, insert a two-pole 40-ampere circuit breaker. Nearly every manufacturer has a different type breaker, so you will have to make sure you have purchased one that will fit your panel. The installation of the circuit breakers will vary also, but in general locate two unused prongs on the bus bar of the panel. Insert the end of the breaker towards the outside first. Then, making sure that the notches on the breaker are lined up with the prongs on the bus bars, press the other side of the breaker down firmly.

Remove one of the knockouts from the panel which will most directly accept the cable, insert a 1-inch NM cable connector in the knockout. Then run the cable through the connector opening until approximately a half-inch of sheathing has passed the connector. Tighten the connector and you're ready to start making connections inside of the panel.

Start with the grounding conductor first. This will be the wire with the green insulation. Strip about a half-inch of insulation from the end of the wire. Then route it neatly around the panel to the grounding or neutral block. Insert it under one of the empty screw terminals. If an individual grounding terminal block is provided in your panel you can tell it immediately by other bare or green insulated wires connected to it. If not, use the neutral terminal block (the one with all the white insulated wires attached). Next, secure the white wire in the cable to the neutral terminal block in the panel.

With these wires in place route the black wire neatly to the new circuit breaker. Insert the wire under one of the screw terminals and tighten the screw. Now repeat this operation with the red insulated wire. Make sure all of the contacts are tight, and replace the panel cover.

Turn the main breaker back on, then turn the new 40-ampere breaker to the on position. You should now be ready to cook.

While we're on the subject of panelboards let me again stress the point of neatness when routing and connecting wires inside of it. To illustrate the panel in Fig. 9-4 is nothing but a mess, while the wiring in the panel in Fig. 9-5 is very neat and shows good workmanship. Which panel would you rather troubleshoot?

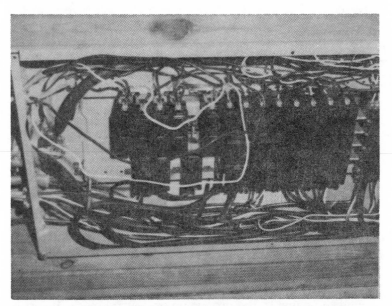

Fig. 9-4. Panelboard showing messy wiring that is very unworkmanlike.

130

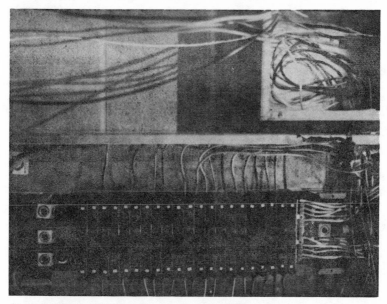
Fig. 9-5. Panelboard wired in a good workmanlike manner.

If the range in question was of the built-in type requiring only a connection to a junction box the range end of the cable will be connected similar to the method described for the range receptacle except that the terminals are in a junction box on the back of the range instead of in the receptacle.

In the case of built-in ovens and cooktops usually a piece of BX cable is supplied with the units. For these units, you will have to install an additional junction box and insert the feeder cable and the cable from the unit in it to make the necessary splices.

Any other type of 240-volt appliances or equipment will be wired in a similar way. First determine the size wire required. Then determine the wiring method and the number of wires required. Run the cable. Next make the connections at the equipment and at the panel as previously described, taking all precautions against electrical shock. Finally make all terminals and splices secure. There, you have saved money by installing your own 240-volt circuit.

Where a plug is required at the equipment the nameplate and the type of cord will dictate the type to install. Table II-2 in Appendix II will help you to select the proper type.

# Chapter 10

# Update Your

# Own Electric Service

All homes and farm buildings that contain an electrical system require an electric service, which is usually supplied by the local power company. This electric service is usually defined as the overhead or underground wires, through which electric service is supplied, between the power company's distribution facilities and the point of connection to the owner's service entrance. Figure 2-4 shows the components of an overhead service drop, that is, the high voltage lines to the last power company pole, the transformer that reduces the high voltage to a voltage usable in the home and the wires from the transformer to the service facilities located at the building or other support.

All components between the point of termination of the overhead service drop and the building main-disconnecting switch, with the exception of the power company's metering equipment, are known as the service entrance, and the wires used in this section of the electrical service are called the service-entrance conductors.

Service-entrance equipment, such as the main switch and circuit breaker panelboards, provides overcurrent protection to the service-entrance conductors and branch circuits and also provides a means of disconnecting the feeders from energized service wires. The electric meter, which is a part of the service-entrance equipment, provides a means of measuring the amount of energy used.

The type of service available depends upon the location, character and size of the homeowner's electrical load. However, the standard service for homeowners and farmers to furnish power to lighting, appliances, and motors up to 5 horsepower, is the single-phase three-wire 120/240 volt service.

If motors larger than 5 horsepower are to be used the local power company should be consulted because motors larger than 5 horsepower have starting characteristics that make their use on most single-phase lines prohibitive. Whenever motors larger than 5 horsepower are used the power company will probably require that a three-phase service be supplied. This can either be a three-phase four-wire 120/208 volt (Y-connected) service or a three-phase four-wire 120/240 volt (delta-connected) service. Of the two, the latter usually is more satisfactory for home and farm equipment loads.

## CORRECT SIZE OF ELECTRIC SERVICE

In general a residence or farm served with electric power can be supplied through only one set of service-entrance wires, except under certain conditions.

The service-entrance wires must have adequate capacity to safely conduct the current for the loads supplied without a temperature rise that may harm the insulation or covering of the wires. They must also have adequate mechanical strength and cannot be smaller than AWG #8 copper or AWG #6 aluminum.

For residential electrical services, the National Electrical Codes gives two methods for determining the size of the service-entrance conductors. One is called the *standard* method and the other is called the *alternate* method.

Using the floor plan of the residence in Fig. 10-1 as a guide the following is an example of employing the standard method of calculating the proper size service to use:

1. Determine the total number of square feet within the building by either measuring the building or by

scaling a drawing of the building. In doing so the building should be measured from the outside walls of the building.

2. The total area in square feet is then multiplied by 3 watts to determine the general lighting load.

3. Each 120-volt small appliance circuit is listed and each is multiplied by 1500 watts.

4. A demand factor may then be applied to the sum of the calculated lighting load and the small appliance load; that is, the first 3000 watts is figured at 100 percent and the remaining wattage at 35 percent.

5. After the demand factor is applied this gives the net load excluding the electric range, water heater and other major appliances.

6. Add 8000 watts for an electric range that is not over 12,000 watts in total output. Refer to the National Electric Code (Article 220) for ratings over 12 kW.

7. Next, add together the nameplate rating (in watts) of all other fixed appliances served by individual circuits not previously accounted for in the calculation. If two major appliance loads are known not to operate simultaneously, e.g., air conditioning and electric heat, only the larger load of the two need be added in the calculation.

8. Total all of the previously calculated loads in step 5 through 7.

9. Divide this grand total load in watts by the line-to-line voltage to obtain the required ampere rating of the service wires and related equipment. For most residential services, the line-to-line voltage will be between 230 and 240 volts.

Therefore, in the residence in Fig. 10-1, the dimensions are 30.5 feet wide by 48 feet long. The total area is found by

$$30.5 \times 48 = 1,464 \text{ square feet}$$

However, a 11.3- by 19.5-foot carport is included in this calculation and should be omitted since it is not a part of the house interior. So 220.35 square feet (11.3 by 19.5) taken away from the original calculation leaves 1,244 square feet of living space.

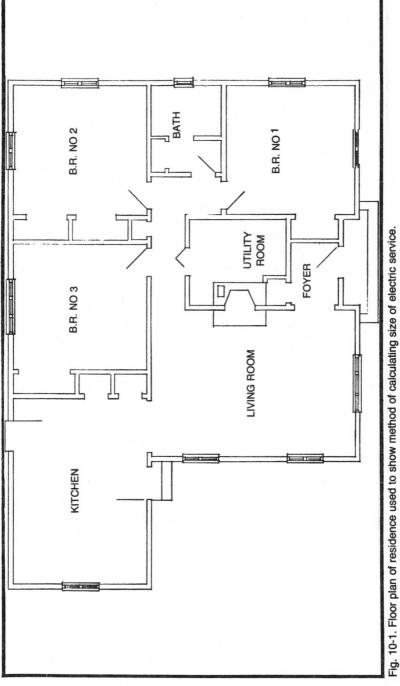

Fig. 10-1. Floor plan of residence used to show method of calculating size of electric service.

The calculations of the lighting load, appliance loads, electric range, and other build-in appliances are shown in the following summation:

## General Lighting Load

| Item | Watts |
|---|---|
| 1,244 square feet at 3 watts per foot ....................... | 3732 |

## Appliance Circuits

| | |
|---|---|
| Kitchen, 3 at 1500 watts per circuit ....................... | 4500 |
| Laundry, 1 at 1500 watts per circuit ...................... | 1500 |
| Total........................................................ | 9732 |

## Application of Demand Factor

| | |
|---|---|
| 3000 watts at 100 percent .................................. | 3000 |
| 6732 watts at 35 percent.................................... | 2356 |
| Net load without major appliances......................... | 5356 |
| Range ...................................................... | 8000 |
| Space heating............................................... | 9400 |
| Clothes dryer............................................... | 4500 |
| Water heater................................................ | 4500 |
| Total load .................................................. | 31,756 |

Then to calculate the required ampere rating divide the total load in watts by the line-to-line voltage. This is,

$$\text{Amperes} = \frac{31,756 \text{ watts}}{240 \text{ volts}} = 132.31$$

The total calculated service-entrance wires and related service equipment should then be a minimum of 133 amperes. But after looking through equipment catalogs, you will find that most service-entrance equipment are rated in 100, 150, 200, etc. amperes, and no conventional fuse or circuit breaker is rated for exactly 133 amperes. Therefore, we will go to the next higher size service which is 150 amperes. If much future expansion is contemplated, it would be better to install a 200-ampere service rather than the 150-ampere type. The additional expense will not be that great and you will have lots of room for expansion and adding additional electrical appliances and circuits in the future.

## INSTALLATION OF SERVICE EQUIPMENT

The drawings in Fig. 10-2 show two types of 200-ampere overhead service entrances. The top drawing shows the service equipment mounted against one end of the house, while the lower drawing shows a rigid conduit mast running through the roof of the house in order to obtain sufficient height for the service drop from the power company's pole.

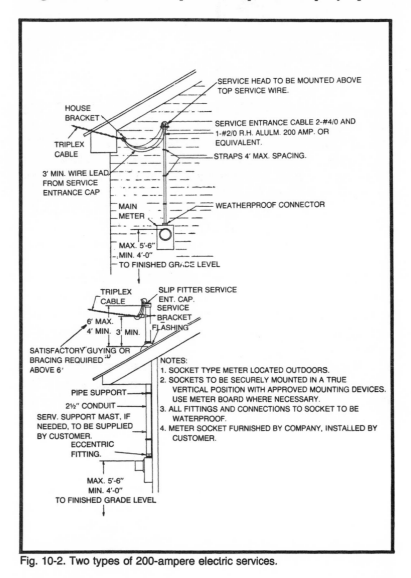

Fig. 10-2. Two types of 200-ampere electric services.

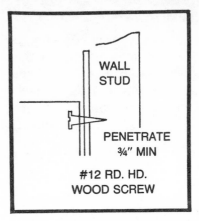

WALL
STUD

PENETRATE
¾" MIN

#12 RD. HD.
WOOD SCREW

Fig. 10-3. Wood screws are used to secure meter base to wood surfaces.

To begin obtain a 200-ampere meter base from your local power company. Usually they are free of charge. At the same time you might take along a sketch of your site plan showing the relationship of your house to existing electric lines and ask the power company the best place (on your house) to install the meter and service-entrance equipment so that the routing of the power company's line will be the most direct without having to cut tree limbs, etc.

Once the location of the service equipment has been determined, remove the front cover on the meter base and mount it to the side of your house a minimum of 4 feet above the ground and a maximum of 5 feet, 6 inches. These heights makes it more convenient for the power company's personnel to read the meter to determine the amount of current you have used over a given period.

If the meter is to be installed against wooden boards, regular wood screws (Fig. 10-3) will be fine; if installed against a masonry wall, then anchors such as the ones in Fig. 10-4 will have to be used. Use a level to make certain the meter base top is level.

Next determine the amount of service-entrance cable you will need from the meter base to the service head. You should leave approximately 3 feet at the service head in order for the power company to make their taps, about 18 inches at the meter base to make the connections inside of it and enough between the meter base and the service head to comply with the following:

- Not less than 3 feet from windows, doors, porches or similar locations.
- Not less than 2 feet above or below telephone wires.
- Not less than 18 feet above public street or roads.
- Not less than 15 feet above residential driveways.
- Not less than 12 feet above finished grade level at any point.

With the length of service-entrance cable determined you will now need the following materials to complete the service from the point of attachment to the building to the meter base: one service head (for 200-ampere service wires), enough 200-ampere service-entrance cable as determined previously, cable straps to secure the cable every 4 feet, enough fasteners to secure the service head and cable straps to the wall surface and a weather-proof service-entrance cable connector to accommodate the termination of the cable at the meter base.

Begin the operation by measuring back from one end of the cable approximately 38 inches. Mark the cable, then strip off the sheathing from that point back to the end of the cable. You will then see two insulated conductors enclosed inside of a bare wire braided around them. Unwind the strands of this bare wire from around the two conductors (Fig. 10-5). Then rewind these bare strands to form a third wire as shown in Fig. 10-6. This is the neutral wire.

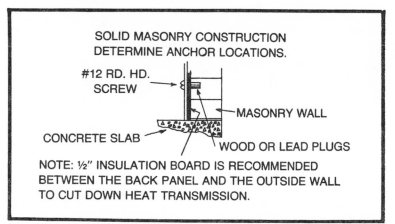

SOLID MASONRY CONSTRUCTION
DETERMINE ANCHOR LOCATIONS.

#12 RD. HD. SCREW

MASONRY WALL

CONCRETE SLAB

WOOD OR LEAD PLUGS

NOTE: ½″ INSULATION BOARD IS RECOMMENDED BETWEEN THE BACK PANEL AND THE OUTSIDE WALL TO CUT DOWN HEAT TRANSMISSION.

Fig. 10-4. Lead or plastic anchors are used to secure meter base to masonry surfaces.

Fig. 10-5. Unwinding the strands of bare wire from around the two insulated conductors.

Now, with the neutral wire between the two insulated wires, bend each of the two insulated wires down along the side of the service-entrance cable as shown in Fig. 10-7. Loosen the clamp on the service head, slip the entire assembly over the service-entrance cable and tighten clamp. You are now ready to install the service-entrance cable.

Locate, by measuring or with a plumb bob, a point directly over the opening in the top of the meter base at the correct height above the finished grade. Mark it with a pencil. Then install an appropriate anchor or fastener at this point through the hanging opening on the service head. Tighten to hold the service head and the top of the cable in place. Then continue down from the service head, securing the cable with cable straps and fasteners not more than every 4 feet. Try to space these at equal intervals for a neat appearance.

Next strip enough sheathing from the opposite end of the service-entrance cable to make the connectors in the meter base. Then take the service-entrance cable connector apart (most come in three pieces) and slip the top part of the clamp on the cable and also the rubber center part. The lower part of

140

Fig. 10-6. When twisted together the bare strands form a third conductor.

Fig. 10-7. Final bending of service entrance conductors prior to installing weatherhead.

141

Fig. 10-8. Service-entrance cable enters meter base by means of a service-entrance connector.

the clamp will screw into the lug on the meter base. Make sure this is tight.

The next step is to insert the three wires from the stripped end of the cable into the hub opening and into the meter base. Then slide the two connector parts (on the cable) down until they come into correct contact with the lower part of the connector screwed into the meter's hub. Tighten the two screws to firmly secure the cable connector in place before applying the weatherproofing compound, furnished with the connector, to the openings where the cable passes through the connector. The cable at the meter base should now look like the one shown in Fig. 10-8.

Make the connections inside of the meter base as shown in Fig. 10-9. This completes the line side of the service entrance. Now for the conductors on the load side from the meter base to the main switch or main panelboard.

The service wires from the meter base to the main disconnect switch or panelboard with a main circuit breaker may also be service-entrance cable, or wires run in rigid or EMT conduit. The main restriction is that the Code states that the service wires must terminate into a main switch or panelboard *immediately* after entering the interior of the house. Authorities differ on the meaning of immediately but few will allow more than 6 feet of unfused conductor inside of the house. If it is not practical to run the service wires immediately from the meter base to the main switch inside the service wires will have to be run on the outside of the house (Fig. 10-10) until it reaches a point where the wires can be run directly into the main fusible service switch inside. Or the conductors must be fused outside at the meter base. Then a four-wire cable can be run from the point of the fuses to anywhere within the house.

Once the cable from the meter base to the service switch is installed, they are connected to the terminals of the switch as shown in Fig. 10-11. Note that the black wire is connected to one terminal of the hot bus bars; the red wire to the

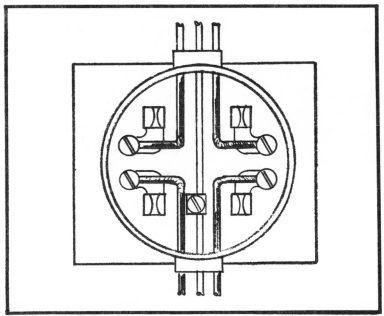

Fig. 10-9. Wiring connections inside of meter base.

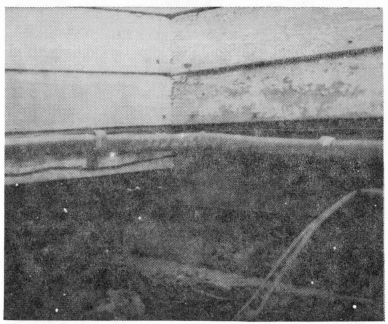

Fig. 10-10. Service-entrance cable run on outside of building when meter base is some distance from the main switch or panel.

terminal of the other hot bus bar; and the white (neutral) wire to the solid neutral block. Also note that AWG #4 bare copper wire is run from the solid neutral block to the closest cold water pipe for the purpose of grounding the equipment. The subject of grounding will be further covered in Chapter 11.

The details of installing an underground 200-ampere service are shown in Fig. 10-12. The meter base is mounted the same as for the overhead service previously described, except that the top hub in the meter base is sealed to keep water out, and the cable or conduit enters from the bottom.

Normally, the homeowner is required to furnish an empty conduit from the meter base to approximately 2½ feet below grade. A 90-degree conduit elbow with bushing is installed on the bottom of this conduit to accept the power company's service wires more readily. The wires (both line and load side) are connected inside of the meter base as described in Fig. 10-12, and the remaining wiring is done in the same way as described for the overhead service.

144

## POLE METERING

In selecting the location of the service entrance and metering equipment on the farm, the electric service should be located as near as possible to the area using the greatest load. In many cases, especially in farmhouses using electric heat, this area will be the farmhouse itself. On the other hand if there is a workshop or other area where many electric motors are used to run machinery, the greatest load could be elsewhere.

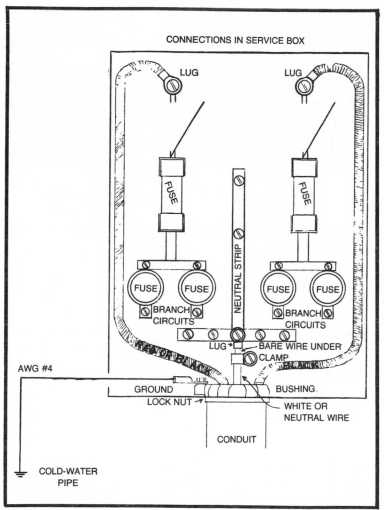

Fig. 10-11. Connections of service conductors inside of main switch.

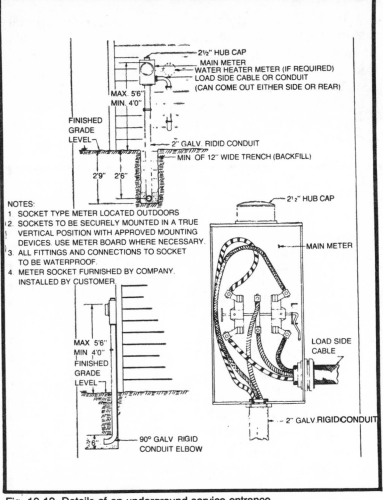

Fig. 10-12. Details of an underground service entrance.

A typical metering pole for farm use is shown in Fig. 10-13. The equipment mounted on this pole includes the power company's 7200-volt line, a transformer to reduce the high voltage to a usable 120/240-volt single-phase service, weatherhead and conduit containing wires to the meter and back up the pole for distribution to other points throughout the farm and a pole-mounted light which is often a dusk-to-dawn mercury-vapor type. Such a service should be located approximately at the center of the greatest electrical load as shown in

Fig. 10-14. In doing so the voltage drop is kept to a minimum, smaller wire sizes will carry the loads, a saving in wire cost, and shorter runs of wires to the various buildings will be required, another savings in wire cost.

A central pole metering is best used when one or more of the following conditions exist:

- A substantial electrical load is installed or contemplated in each of two or more buildings
- Buildings containing electrical loads are scattered

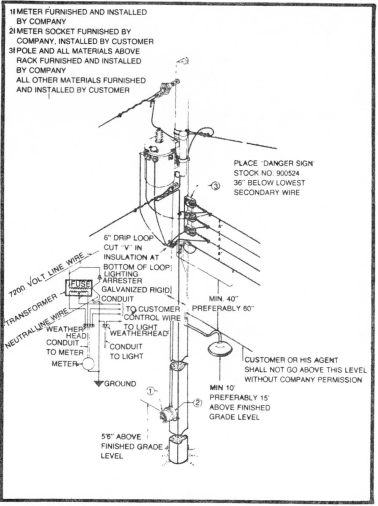

Fig. 10-13. A typical metering pole for farm use.

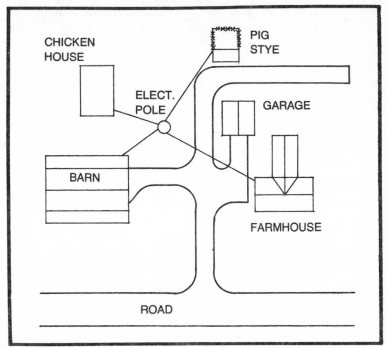

Fig. 10-14. Site plan showing location of metering pole.

■ Due to trees or other obstacles outbuildings cannot easily be fed except by a pole installed to miss these obstacles

While we are on the topic of central pole metering, it may be well to include the main advantages of such an electrical installation.

First, less investment in electrical wires will be required for feeders from the meter pole to buildings than for most other meter locations since the meter pole can be located as near as possible to the heaviest electrical load on the farm.

When it becomes necessary to change or alter the wiring or service in one of the farm buildings the wiring to other buildings need not be disturbed. Merely disconnect the wire taps to the building where the change is to be made, and leave the other wires energized.

Similarly, an electrical fault like a short circuit causing one of the buildings to be out of commission until the fault can be corrected will not interrupt service to other buildings. Nor will the loss of one building by fire or similar mishap interfere with the electrical service to other buildings.

## PARTIAL UPDATING OF ELECTRIC SERVICE

There will be times when you have found that your service wires are large enough to handle your total electrical load, but you just don't have enough space in your fuse cabinet or panelboard to accommodate additional needed circuits. Or perhaps you find that several circuits in your home are connected to one fuse or circuit breaker terminal causing them to blow or trip frequently. What do you do then?

Maybe your original service equipment consisted only of a fuse cabinet that contained spaces for an electric range, and four circuits only. Now you want to add an electric water heater and a clothes dryer, but you have no more spaces to add them. Or perhaps your existing service has been added too so often that your service equipment looks like the photo in Fig. 10-15. In any case the solution means an updating of your service equipment.

First, make a survey of the existing circuits in your home. Then, with your family, determine what additional appliances you expect to have in the near future. After deciding upon the number of circuits required, add about twenty

Fig. 10-15. Service equipment that has been added during the years.

percent to this number to handle any additional circuits you have overlooked or that you do not know about at the time.

If, after making the calculations described previously on how to size your service wires, you find that your present size is adequate purchase only a new circuit breaker load center containing the required number of spaces for circuit breakers that will provide over-current protection for the number of circuits that you will install.

Acquire all of the necessary materials first. Double-check with an electrician or power company representative to make sure. Then, with all the necessary material on hand, perform all of the work you can without having to disconnect the power. This might include mounting the new circuit breaker panel beside the existing piece of equipment and installing any new circuits and connecting them to the new panel. Of course they will be dead until the new panel is actually connected to your existing service wires. You can even install a new ground wire (as described in Chapter 11) before you make the actual changeover.

Fig. 10-16. A modern service entrance circuit breaker panel.

When you are sure that you have done all the work possible without disconnecting the power (in order to save time), have the power company pull their meter which will turn off the power from the load side of the meter base to your old service equipment. Then disconnect the service wires from the old switch and reconnect these to your new circuit breaker panel as shown in Fig. 10-11. You should be able to make this change fast enough so that the person from the power company could wait to replace the meter, which in turn, will make your new panel hot. Then transfer the remaining circuits in your old panel to the new; remove the old service equipment entirely, and you will have a neat installation as shown in Fig. 10-16.

# Chapter 11
# Grounding—A Life Saver

All electrical systems must be grounded in a manner prescribed by the National Electrical Code in order to ensure safety of life and property. The main purposes of grounding an electrical system include:

- Limit the voltage from phase to ground on a circuit
- Protect the circuit from exposure to lightning
- Prevent voltage surge higher than that for which the circuit is designed
- Protect the circuit from an increase in the maximum potential to ground due to normal voltage

In order for you to make use of the protective feature of grounding, you must first understand how a ground functions in relationship to an electrical system and how circuits are grounded.

Figure 11-1 shows a fundamental electric service such as might be found in the home or on the farm. At the main panel, where the service-entrance wires enter the house from the power company's meter, the neutral or white wire is thoroughly grounded by means of a connection to a cold-water pipe. If a metal cold-water pipe is not available, the electric service may be grounded to a driven ground rod as shown in Fig. 11-2.

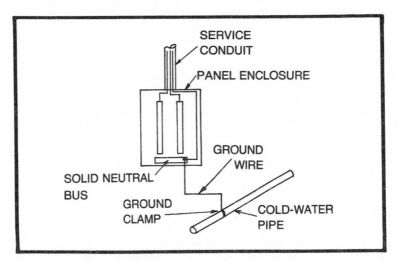

Fig. 11-1. Grounding of an electrical system to a cold water pipe.

The branch circuits running from the main panel to various outlets throughout the home and the metal enclosures of the outlet boxes are considered to be adequately grounded when they are mechanically connected to each other and to the service-entrance equipment metal enclosure. This is accomplished either by a metallic raceway (conduit or armored cable) or by a ground wire enclosed within nonmetallic cable.

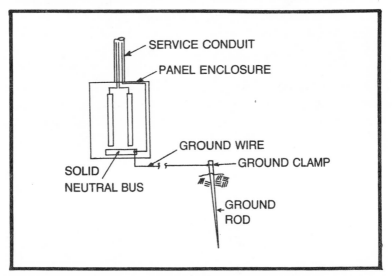

Fig. 11-2. Grounding of an electrical system to a driven ground rod.

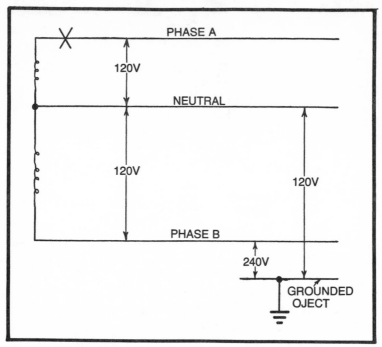

PHASE A

120V

NEUTRAL

120V

120V

PHASE B

240V

GROUNDED
OJECT

Fig. 11-3. Wiring diagram of a typical 120/240 volt electric service with the neutral not grounded.

Now let's see how a system ground, as just described, limits the voltage from phase (one hot wire) to ground. Figure 11-3 shows a wiring diagram of a typical 120/240-volt three-wire electric service. In this diagram the neutral is not grounded. Assume that phase A develops a ground fault as marked by the $X$ on the circuit line and does not blow a fuse. The voltage from phase B to ground will then be 240 volts instead of the normal 120 volts. This means that any grounded 120-volt equipment connected to phase B will receive 240 volts and more than likely damage the equipment. A person who is grounded—for example, standing on a cement basement floor—and comes in contact with phase B will receive a shock from 240 volts instead of 120 volts. On a damp basement floor this could be fatal.

Grounded equipment also provides a direct path to ground from voltage surges outside of the home that come in on the electric lines. Lightning would be one example. In this case the high voltage caused by lightning coming over the wires will be directed to ground rather than coming in on the house wiring. In other words the system ground acts like a lightning arrester since the resistance to ground is less than the resistance of the house wiring.

To better illustrate this fact, we will use water to simulate the flow of electricity. Most free-flowing substances—that is, water and electrical current—always seek the path of the least resistance. Take for example a concrete dam across a stream of water as shown in Fig. 11-4. Normal water level is 2 feet. A 2-inch pipe is placed at the foot of the dam to carry off the normal flow of water to irrigate a crop field near the stream. The soil can easily handle the normal flow of water through this 2-inch pipe without any damage to the crop. Let the 2-inch pipe represent the service wires supplying power to your home, and the crop field represents you home utilizing normal electrical power. The large 10-inch pipe inserted through the dam above the normal water level acts as an

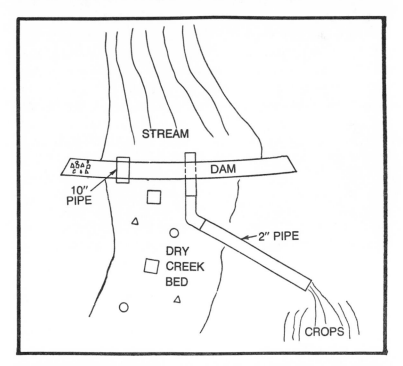

Fig. 11-4. Drawing of a creek dam to show how the system grounds protects the home electric service.

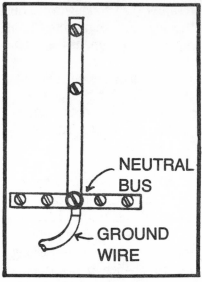

Fig. 11-5. Location of ground wire connection inside of main switch.

NEUTRAL BUS

GROUND WIRE

overflow pipe for high water, and may be compared to the system ground on your home electrical system.

Now suppose a heavy rain storm occurs upstream which causes an enormous amount of water to flow down the stream, much the same as high voltage caused by lightning coming toward your home electrical system over the service wires. Will the water damage the crop or will the higher voltage damage your electrical system? Neither is likely since the 10-inch pipe has less resistance than the 2-inch pipe feeding the irrigation system. Most of the excess water will flow through the 10-inch pipe and be diverted away from the crop field into the stream bed. Likewise, the system ground on your electrical system will divert most of the high voltage away from your interior wiring and carry it directly to ground because the ground has less resistance than the remaining wiring in your home. You should now have a fairly good idea of how a system ground operates.

To insure that your electrical system is properly grounded, the neutral wire of the service-entrance conductors should be grounded either at the meter base or at the main switch enclosure ahead of the main switch as shown in Fig. 11-5. This is accomplished by connecting a grounding conductor to the neutral bus of the panel or meter base and by connecting the other end to a cold-water pipe or other suitable grounding electrode. No further connection should be made to a grounding electrode on the load side of the main-service switch.

You must also make certain that you really have a good ground—not more than 25 ohms resistance. If there is any doubt that the ground resistance is higher than 25 ohms, a ground test should be made with a megger ground tester. Usually the local power company will make this test for you at little or no charge.

The size of the ground wire is also important to ensure a proper ground. Table 11-1 can be used for sizing grounding wires for any residential or farm system ground, the one at the service-entrance location.

Remember if no cold-water pipe is available at the service-entrance location or else the cold-water pipe is made of some other material than metal (PVC plastic for example) the grounding wire must be connected to some other grounding electrode, such as a ground rod shown in Fig. 11-6.

Notice in Fig. 11-6 that a grounding conductor is used to bond the ground rod to the copper cold-water pipe inside of the house, although the main water line from the pump to the house is PVC plastic. The reason for this is that sometimes the water inside of the PVC and copper pipe will offer less resistance to ground than the ground rod, and many heating elements in hot-water heaters have been burned out due to lightning coming in on the water pipes. The additional ground-

Table 11-1. Grounding Wire Sizing

| Largest Service Wire (AWG/MCM) | | Grounding Wire(AWG) | |
|---|---|---|---|
| Copper | Aluminum | Copper | Aluminum |
| 2 | 0 | 8 | 6 |
| 1 | 00 | 6 | 4 |
| 0 | 000 | 6 | 4 |
| 00 | 0000 | 4 | 2 |
| 000 | 250 | 4 | 2 |

ing conductor from the ground rod to the copper cold-water line will help prevent this from happening.

## EQUIPMENT GROUNDING

The system ground just described means that the neutral wire of the service-entrance conductors is grounded. Therefore, the neutral wire is a *grounded* wire or conductor and should have white insulation in all cases. A *grounding* conductor or wire, however, is a wire used to ground circuits and metal portions of electrical equipment used on the circuit. This is referred to as equipment grounding and is also very important on any electrical system. To show just how important this equipment ground is let's look at one actual case that happened in a small Virginia town.

A housewife noticed that her washer in the basement of her home occasionally shocked her when she touched the metal cabinet, and she finally called an electrical contractor to check it. This was a good move on her part, but she didn't wait for the contractor to arrive before she finished her load of clothes. When the contractor entered the basement he found the woman lying on the basement floor, dead from electrocution.

An examination of the washer showed that the metal frame was not grounded as there was no equipment ground attached and it was setting on rubber-padded legs. One of the hot wires inside of the metal cabinet had rubbed against a sharp corner. Vibration had caused the metal cabinet corner to cut into the wire's insulation and come in contact with the bare hot wire. Since the cabinet was not grounded, this did not blow the fuse as it would have had the metal cabinet been properly grounded. Thus, the entire washer cabinet and frame was hot like the wire it had come in contact with.

Perhaps the times before when the woman had been shocked the washer motor was not running which lessened the current received by the shock or maybe the floor was not as wet as it was the last time. Probably, the woman was standing in water splashed on the floor from the washer and touched the washer cabinet while it was running and the combination of the two killed her.

158

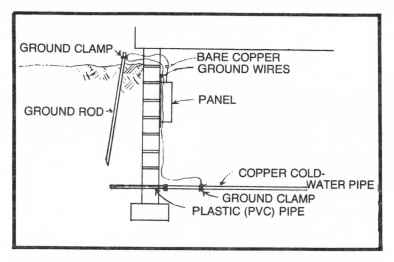

Fig. 11-6. Grounding of a home electric system where a PVC (plastic) pipe is used for the water supply.

More fatal electrical accidents have occurred in the basement and bathrooms of homes than any other area. The reason being water. Yes, water on the floor enormously increases the danger of electricity because the water reduces the resistance of the floor itself causing you to be a good ground. And we have already discussed why electrical current flows along the path of the least resistance. Because of this fact both portable and fixed appliances used in the basement or other damp locations should be especially checked for proper grounding.

If your home has a modern grounded electrical system that third prong on appliance cords (the one that many people cut off) takes care of the job for you. However, if your electrical system is of the older types without an equipment ground run with your branch circuit you will have to ground your appliances by some other means.

For heavy appliances such as an electric washer or dryer, secure a piece of AWG #10 bare copper wire under one of the screws on the metal cabinet. Most of these appliances have a special grounding screw which is plainly marked on the back. The other end of the wire should be attached to the closest cold-water pipe with a ground clamp like the one in Fig. 11-7.

Fig. 11-7. Ground clamp used for connecting ground wire to either a cold water pipe or ground rod.

## PORTABLE APPLIANCES & HAND TOOLS

I'm sure that you have noticed that third prong on cords of metal hand tools and appliances and also the adapter that comes with the cords in case your receptacles are not of the grounding type. Many have found this third prong to be a nuisance in homes without the proper receptacle to plug them in, and have cut this prong off. But in doing so, they eliminated one of the best protective devices against electrical shock. It might be compared to a skydiver leaving his spare chute behind. He probably won't need it, but if he does....

Now let's see just what this third prong on cords has to do with preventing electrical shock. Look at the drawing in Fig. 11-8. The square represents the metal housing of an electric drill motor, the coils are the motor windings and the arrowhead represents the switch on the motor, which is usually a trigger type.

Line A in Fig. 11-8 is the neutral wire and line B is the hot wire. Both of them combined form a 120-volt circuit to the drill motor. Now if a fault developed at point C, that is, the insulation of the hot wire is worn and allowed the bare hot wire to come in contact with the metal frame, a harmful shock could occur when the user plugged the cord into an outlet. He himself would have to be grounded, but if he's wearing leather sole shoes on a basement floor this will do it.

However, a third prong on the plug means that the prong is connected to a grounding conductor inside of the cord. If it is plugged into a grounded receptacle the metal case of the motor will be grounded as shown in Fig. 11-9. Then if a fault occurs such as at point D the line is short-circuited and blows the fuse or trips the circuit breaker. If the fault occurred at point E the fuse would again blow when the switch on the

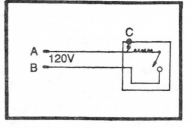

Fig. 11-8. Wiring diagram of a non-grounded drill motor.

drill was closed. If the fault occurred at point C (on the neutral wire) nothing would happen as the frame is already grounded.

A separate grounding wire can easily be added to any tool or appliance if provisions were not made at the factory. The best way is to replace the existing cord with one that has the grounding wire enclosed inside of it along with a grounding type plug. Then you simply connect the white and black wires to the same wires of the tool or appliance, and connect the bare or green grounding wire under any convenient screw on the frame or case. Of course the receptacle has to be the grounded type for the new cord to do any good. If it isn't, then the grounding should be done to a water pipe as described earlier.

If you don't want to invest in another three-prong cord, you can take a piece of bare copper wire, fasten one end under any convenient screw on the frame or case and then run this wire along the outside of your old cord securing it in place with electrical tape. Leave enough on the plug end of the cord to fasten to a ground such as a screw on a nongrounded receptacle when the outlet box is grounded, etc.

Those of you who live on farms should know that livestock are extremely sensitive to electric shock, and voltages that would not affect a human are frequently fatal to cattle and horses. Therefore, all wiring systems and metal frames of

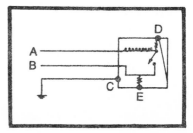

Fig. 11-9. Wiring diagram of a grounded drill motor.

electrical equipment should be properly grounded around the farm.

While we're on the subject of livestock a comment concerning the use of electric fences is in order. Many farmers have made it a practice to install porcelain insulators on fence posts, run bare galvanized wire around the area, and connect the wire to house current—either through an electric lamp or dead-ending the hot wire. While this practice works it has also caused the death of many cows and horses during rainy or damp weather.

For safety reasons, every electric fence should have an electric-fence controller installed to limited the current on the output side to 10 milliamperes. Tests have proved that an animal can safely stand 10 milliamperes of current for a very short time without suffering any ill effects from it. The controller should also have an interrupter to interrupt the current at regular intervals as current as small as 3 milliamperes can be fatal to livestock and humans if not interrupted at regular intervals. Most approved electric-fence controllers also contain fuses to protect them against short circuits or a high-voltage surge which may occur during a lightning storm.

## SUMMARY OF EQUIPMENT GROUNDING

Electric ranges and clothes dryers, when fed with a three-wire 120/240-volt circuit, are grounded by means of the neutral wire. However, an additional grounding wire enclosed inside of the cable and also attached to the frames of these appliances will certainly not hurt anything. For all other appliances, including motors used around the home and farm, a separate gounding wire should be installed. As mentioned previously, this can consist of a metallic conduit or an additional wire enclosed in NM cable. The grounding wire must be run with the circuit wires, and it may be bare or covered with green insulation. The wire can contain no current as can the neutral wire.

All portable and fixed tools operated by electric motors should be equipped with suitable grounding means, either by a grounding wire run with the feeder circuit to permanently located ones or by means of a three-prong appliance cord connected to a grounded duplex receptacle in the case of portable ones.

If this chapter has not impressed upon you the importance of a properly grounded electrical system just pick up any newspaper and read the fatalities reported due to contact with live electrical appliances. The number is sickening when most could have been prevented if the *murder weapon* had only been grounded.

# Chapter 12
# Lighting the Home's Interior

Properly designed and controlled lighting can be one of the greatest comforts and conveniences that any homeowner can enjoy. Electric lighting is also one of the interior decorator's most versatile tools as light has certain characteristics that can be used to change the apparent shape of a room, to create a feeling of separate areas within a room, to create a variety of moods or to alter architectural lines, forms, color, pattern or texture. For these reasons, proper lighting should be considered equally as important as the heating system, the furniture placement, and other items considered necessary for comfort living.

While residential lighting calculations need not be as elaborate as might be required for a school classroom or a commercial application, some guide should be followed to assure that the proper amount of glare-free illumination is obtained in all areas. With the use of such a guide, various possibilities of lighting are available which are limited only by the homeowner's imagination. Obtain and study several of the residential lighting catalogs available free of charge. With your own ideas as to the types of lighting fixtures you desire to use, the following lumens-per-square-foot method will help you determine the number and wattage of lamps for most areas around the home.

The following gives the required lumens per square foot for various areas in the home and also for various locations around the farm.

## Living Quarters Lighting

| AREA | LUMENS/SQ FT |
|---|---|
| Living room | 80 |
| Dining room | 45 |
| Kitchen | 80 |
| Bathroom | 65 |
| Hallway | 45 |
| Laundry | 70 |
| Workbench | 70 |

## Dairy Barn Lighting

| AREA | LUMENS |
|---|---|
| Litter alley | 1010 per two stalls |
| Feed alley | 1010 per 10 ft of wall |
| Loose house barn | 1010 per 150 sq ft |
| Box stalls and pens | 870 per pen |
| Milking room | 3240 per cow rear |
| Milk house | 45 per sq ft |

## Poultry House Lighting

| AREA | LUMENS |
|---|---|
| Laying house: | |
| Dim evening lights | 1750 per 400 sq ft |
| Bright evening light | 1750 per 200 sq ft |
| All-night lights | 10 per sq ft |
| Brooder house | 45 per sq ft |
| Egg storage and handling: | |
| General | 10 per sq ft |
| Work area | 3250 |
| Cleaning and dressing | 70 per sq ft |

## Crop Storage Barn

| AREA | LUMENS |
|---|---|
| Feed grinding | 23 per sq ft |
| Feed storage | 8.75 per sq ft |

Barn floor.................................8.75 per sq ft
Corn crib..................................12 per sq ft
Hay loft...................................7 per sq ft
Silo:
        Top of chute.....................300-watt lamp
        Ceiling...........................150-watt lamp
Potato storage, etc.......................12 per sq ft
Fruit/vegetable storage..................12 per sq ft
Greehhouse...............................8 per sq ft

In using this method, it is important to remember that lighter room colors reflect light and darker colors absorb light. The listings contained herein are based on rooms with light colors so if your room surfaces are dark, the total lumens obtained in your calculations should then be multiplied by a factor of 1.25 to insure the proper amount of illumination. These listings are further based on surface-mounted lighting fixtures; if the area to be lighted contains recessed fixtures, the total lumens obtained from the recessed fixtures should be multiplied by 0.60.

Now for a typical application of the data.

The floor plan of a residential living room is shown in Fig. 12-1. From the listing above we can see that 80 lumens per square foot would be required to obtained the recommended level of illumination. The amount of lamps to obtain this level must be found next.

Scale the drawings to find the dimensions of the room. Or in the case of an existing residence, actual measurement with a tape measure should be performed. In doing so, we find that the area of the living room in Fig. 12-1 is (13.75 feet by 19 feet) 261.25 or 261 square feet. The area is then multiplied by the required lumens per square feet or $261 \times 80 = 20,880$. This is total lumens required in this area to obtain the proper illumination level.

The next step is to refer to manufacturer's lamp data (see Appendix IV) to select lamps that will give the required amount of lumens. At the same time you should be looking through residential lighting fixture catalogs to get some idea of the type of lighting fixtures you plan to use in the area, as well as the location of the fixtures.

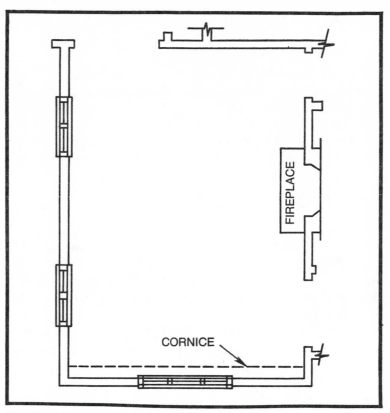

Fig. 12-1. Floor plan of a typical residential living room.

Let's assume that we admire the stone fireplace in the area and would like to highlight the texture of the stones. In looking through a residential lighting fixture catalog we notice a type of recessed fixture which is intended to *wash* walls with light. This seems to be perfect to light the stone fireplace and chimney. So we read the description of the fixture—each will contain up to one 150-watt lamp. Referring to the lamp data in Appendix IV, we see that a 150-watt inside frosted lamp gives off approximately 2880 initial lumens. Since we are using two of these lamps we now have selected lamps that give a total of 5760 lumens. This means that we have to select lamps for 15,120 more lumens (20,880 total required lumens minus 5760 lumens just selected = 15,120 lumens).

Oops! We forgot that these fixtures are recessed. Therefore, we must multiply the total lumen output of the two fixtures by a factor of 0.60, which gives us only 3466 total lumens for the two fixtures. This means we now have 17,414 more lumens to account for.

Your spouse had previously expressed a desire for some indirect lighting in this area so you have decided to use a drapery cornice along the entire front wall of the living room. You then selected four 40-watt fluorescent bare strip fixtures to be concealed behind the cornice. Each fixture will contain one 40-watt warm-white fluorescent lamp rated at 2080 lumens; this gives a total of 8320 lumens for the four fixtures. Only 9094 lumens left to account for.

Two three-way (100-, 200-, 300-watt) lamps in table lamps will be used on end tables located on each end of a sofa. Two of them will give a total of 9460 lumens when switched to the high rating. When combined with the other lamps in the area this gives a total lumen output of 21,246 lumens, which is close enough to our recommended figure of 20,880 lumens to be nearly perfect.

As a final touch, dimmers should be added to the recessed fixtures at the fireplace and perhaps on the cornice fluorescent light. Then, since the two three-way lamps can also be dimmed by switching to different wattages, you can vary the living room's lighting levels exactly to the activities—low for a relaxed mood, bright for a gay, party mood.

Obviously this method of calculating residential lighting requirements makes it possible to quickly and accurately determine the number and size of light sources to achieve the recommended lighting level in any area of the home.

## SELECTING LIGHTING FIXTURES

Since there are numerous variable factors involved in selecting the type of lighting fixture for a given application, no set rules are available. However, the following guidelines should prove useful to home owners:

- Determine the total lumens required for a given area from the previous listing in this chapter and by methods demonstrated herein

- Study residential lighting fixture catalogs to see what types of fixtures are available. Also study interior decorating magazines for ideas
- Prepare a master plan of the area. Then select a fixture or fixtures that fits into the architectural or decorating scheme of the area
- When selecting the fixtures read the manufacturer's description in the catalog to find out the number and size of lamps recommended for the fixture you have selected. Then look in Appendix IV of this book to obtain the lumen output of the lamps

Here are some additional hints on selecting proper lighting fixtures for various areas in the home. For added convenience, the separate areas are broken down so that each may be discussed individually.

## Living Room

This is the area in the home where guests are entertained and where the family gathers to relax, watch TV, or engage in conversation. Lighting in this area should emphasize any special architectural features such as planters, bookcases, fireplaces, etc. Pull-down lighting fixtures or table lamps placed at chairs or at the ends of sofas are used for reading.

## Dining Room & Kitchen

In residential dining rooms a chandelier mounted directly above the dining table and controlled by a dimmer/switch becomes the centerpiece of the room while providing general illumination. The dimmer adds versatility to the general illumination in that the lighting can be dimmed for formal dining or made bright for an evening of cards.

In addition to the center chandelier supplementary lighting at the buffet and sideboards are often desirable. Use recessed accent light for a contemporary design and wall brackets to match the chandelier for a traditional setting. Other possibilities consist of using concealed fluorescent lighting in valances or cornices.

The ideal general lighting system for a residential kitchen would be a luminous ceiling. This lighting arrangement gives a skylight effect, but it is also the most expensive to install. The effect is achieved by installing rows of bare fluorescent strip lighting fixtures above a dropped ceiling, consisting of ceiling panels with attractive diffuser patterns. The fixtures should be spaced approximately 2 feet on center.

If the luminous ceiling is not employed a fixture mounted in the center of the kitchen area will provide general illumination. Additional lights should be mounted over the sink, electric range and under wall cabinets to provide light down on countertops.

Many people spend at least a third of their time in their bedroom, and the quality and layout of the bedroom lighting should reflect this fact.

Basically bedroom lighting should be both decorative and functional with flexibility of control in order to create the desired lighting environment. For example, both reading and sewing are two common activities occurring in the bedroom, and both require good illumination to lessen eye strain. Other activities, however, such as casual conversation, lovemaking or watching TV, require only general nonglaring room illumination, preferably controlled by a dimmer/switch.

Proper lighting in and around the closet area can do much to help in the selection and appearance of clothing, and supplementary lighting around the vanity will aid in personal grooming.

Good light is needed in the bathroom for good grooming and hygiene practices. If the bath is small usually the mirror light combined with a tub or shower light will suffice. On the other hand if the bathroom is large a bright central light source is recommended with supplemental light at the mirror and similar areas.

## Basement, Utility Room & Workshop

Basement, utility room and workshop require a similar amount of illumination and lighting techniques. The general lighting need only be about 45 lumens per square foot in these areas, but supplemental light over work areas of at least 70 lumens per square foot should be provided. Normally, these areas will not be visited by guests, so you can use inexpensive fixtures in these areas, and use the money saved to improve the lighting in other areas of the home.

A well-designed lighting layout for a family room would include graceful blending of general lighting to illuminate the overall area with well-chosen supplemental lighting to illuminate certain individual seeing tasks. For example, diffused recessed incandescent lighting fixtures installed flush with the ceiling of the family room will furnish even glare-free light throughout the room if the proper number are installed and they are spaced correctly.

Lamps concealed behind cornices near the ceiling will enrich the natural beauty of paneled walls or the texture of brick or natural stone walls. This technique is also very effective over bookshelves where the light is positioned to shine down over books with colorful bindings. Fluorescent lamps concealed in a cove lighting system will not only furnish excellent indirect general illumination for a family room, but will also give the impression of a higher ceiling. This is a very desirable effect in low-ceiling family rooms in basements of homes.

We have only touched upon the many possibilities, but one thing to keep in mind is that the lighting layout for any family room should be highly flexible since this area is used for a variety of daily activities. For instance, casual conversation is enhanced amid subdued, complexion-flattering light such as incandescent or warm-white fluorescent lamps controlled by a dimmer switch. Game participants feel more comfortable in a uniformly lighted room with some additional glare-free light directed onto the playing areas. Low-level lighting over the bar area should be just bright enough for mixing a drink or having a snack. TV viewing requires only softly lighted surroundings, while reading requires a somewhat brighter light source with light directed on the printed pages. By now you should be getting some ideas of your own, so let's see how we can actually install some of these lighting fixtures.

## INSTALLATION OF LIGHTING FIXTURES

Once you have determined the lighting layout and the type of lighting fixture your next step is to install the fixture outlet, runs wires to it, provide a switch or other means of

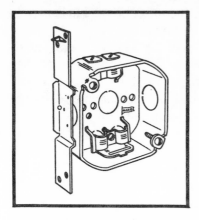

Fig. 12-2. Four inch octagonal box on which most lighting fixtures are mounted.

controlling it and finally install the fixture itself. This is actually the easiest part of proper lighting. The design and selection of the fixture are the most difficult steps.

Actually the wiring of a lighting fixture outlet is no different that the wiring for a duplex receptacle as described in Chapter 6, or adding a wall switch as discussed in Chapter 7. The only difference, and this is minor, is possibly the type of outlet box.

Most lighting fixtures are designed to be mounted on a 4-inch octagon box such as the one shown in Fig. 12-2. This figure shows a 4-inch octagon box with NM cable clamps and also a bracket for mounting in unfinished areas. The bracket is merely nailed or secured by some other means between two ceiling joists.

For existing finished areas where the box opening must be cut into the finished ceiling, an octagon box with mounting ears (Fig. 12-3) should be used or else a regular box secured

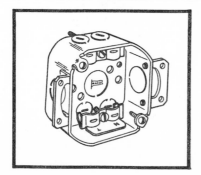

Fig. 12-3. Octagonal box with mounting ears used in old work.

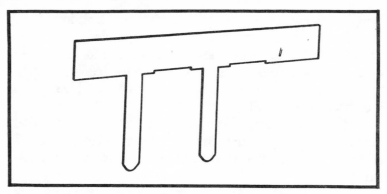

Fig. 12-4. Box supports used to secure outlet boxes in old work.

by switch box supports as shown in Fig. 12-4. However, neither of these supports can take very much weight. If a heavy chandelier is to be installed the outlet box should be firmly secured to a structural member even if cutting and patching of the ceiling is necessary.

A 120-volt lighting fixture will require two wires to feed it, just like any other two-wire electrical device. If the fixture

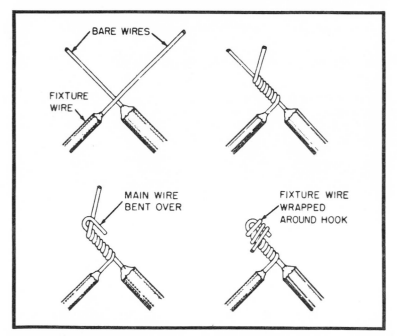

Fig. 12-5. Necessary steps to make a good fixture joint.

contains its own means of control, such as a built-in switch, the wires are run exactly as described for duplex receptacles in Chapter 6. The wiring methods described for either the receptacle or switch circuits will be fine for feeding lighting fixtures and methods of running the circuits are identical to the receptacle or switch wiring.

If the newly installed lighting fixture requires an external means of control use the techniques described in Chapters 7 and 8 to arrive at the best means of control for any given situation.

Once the outlet and relative wiring is installed you simply install the fixture according to the easy-to-read instructions enclosed in the packing carton of nearly every fixture. All will have two wires, and some a third bare grounding wire. The grounding wire should be secured to a proven ground, such as the outlet box, under a screw or by using an approved ground clip. The remaining two leads are connected to the feeder leads—black to black, and white to white. Make either a rattail or fixture-wire splice (Fig. 12-5), and secure with wire nuts.

# Chapter 13
# Outdoor Lighting

Outside lighting should be a very important consideration for the farmer or homeowner. Properly selected, located and installed outside lighting can provide a safeguard against accidents at night, increase the afterhours farm efficiency and protect the home and farm from prowlers. It also welcomes guests and lights their way to the house entrance. It creates a hospitable look and turns the area surrounding the home and farm into an extra living and play area during warm weather. In addition lighting reveals the beauty of flower gardens, trees and foliage; expands the hospitality and comfort of patios and porches; and stretches the hours for outdoor recreation or work. In fact the time required to do after-dark work around the farm such as watering, feeding and housing livestock, as well as storing equipment and similar jobs, can be greatly reduced with the aid of adequate light.

The yard lighting around the farm should be so located that the most frequently used walkways and work areas are well lighted. These same lights should be controlled from the buildings most accessible to the area or else mounted to poles to which the lighting fixtures and controls can be mounted.

Outdoor lighting between buildings should be so controlled that the light may be switched at either building. As described in Chapter 8, such switching requires the use of three- and four-way switches. If there are several such locations or long distances between the lights and the switches it may be better to use low-voltage remote-control switching to cut down on the amount and size of wire, which, in turn, will save the farmer money and time.

Some farmers who use low-voltage control for their outdoor lighting also have a master switch installed in the farm residence, either in the bedroom or kitchen, or both. This is really an economical investment as a protection against prowlers. With lights installed at poultry houses, feed storage buildings, and other outbuildings the farmer may turn any or all of the lights on from the house should he suspect any disturbances from prowlers.

Other farmers, as a matter of added insurance, use dusk-to-dawn lights controlled by photoelectric switches. The lights are automatically turned on at dark and off at daybreak. Most power companies even have a policy where they will install dusk-to-dawn lights on the farmer's property, then furnish power for them, all at a predetermined flat rate.

Figure 13-1 shows a dusk-to-dawn light mounted on the farm's metering pole. Since the pole is located approximately in the center of the farm buildings this one light suffices for the entire farm lot. At the present time, the monthly lease for this light is still low but increasing all of the time. It may be cheaper to purchase one of these fixtures, available at most department stores, and do the installation yourself. In some locations the cost of these flat-rated dusk-to-dawn arrangements from the power company has doubled since initially installed.

## INSTALLING OUTDOOR FARM LIGHTING

Basically there are three methods of installing circuits for outdoor lighting: in metallic or PVC plastic conduit with weatherproof fittings; underground wiring using type UF cable; or overhead wiring using single-conductor wire. All are satisfactory for outdoor wiring, but there are certain cases where one type may be preferred over another. The following paragraphs will explain the advantages and disadvantages of each of the wiring methods mentioned.

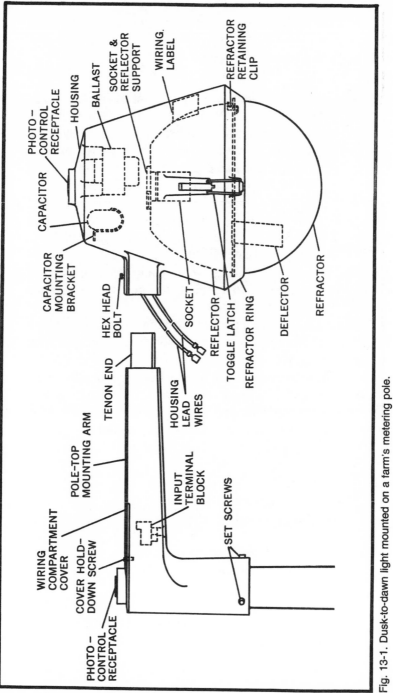

Fig. 13-1. Dusk-to-dawn light mounted on a farm's metering pole.

177

In areas where the soil is not too rocky, underground wiring should be the preferred method as it has the following advantages over the other types:

- Underground lines are not subject to damage as are overhead lines. Overhead lines are sometimes struck by trucks passing beneath them; they often break from heavy ice loads during winter months and repairs and maintenance on underground lines are much easier than on overhead lines.
- Underground lines, of course, cannot be seen and therefore presents a better appearance of the farm yard. A lot of overhead lines give a cluttered look to the most well-kept farm.
- Underground wiring, when cable is used, is often less expensive to install than other tpes of outdoor wiring.

Figure 13-2 shows a section through a farmhouse and a typical outbuilding requiring electric service. In this instance the main-service panelboard is located in the basement of the house, and it is desired to run an electric feeder to the barn approximately 300 feet away. We begin the design by determining the estimated load that will be connected. The following summary shows this farmer's requirements:

| | | |
|---|---|---|
| 12 | 150-watt lamps | 1800 watts |
| 4 | 120-volt duplex receptacle circuit with maximum of 1000-watts operating on circuit at any time | 1000 |
| 1 | 240-volt, 20-ampere receptacle for various portable heaters, etc. | 3800 |
| | Total load | 6600 watts |

Since a three-wire 240-volt cable will be used the total amperage may be found by the formula,

$$\text{Amperes} = \frac{6600 \ (\text{watts})}{240 \ (\text{volts})} = 27.5$$

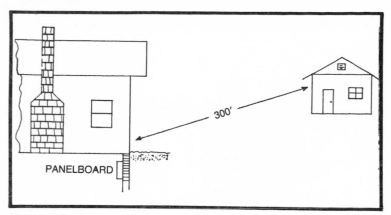

Fig. 13-2. Cross section through a farmhouse and an outbuilding.

We next determine the wire size by considering the voltage drop for the distance of 300 feet. Refer to Table 9-2 to obtain the recommended wire size to carry the estimated load at a voltage drop not to exceed three percent. Look down the left-hand column until you come to 30 amperes. Then read across the line to the column under 300 feet. The table gives AWG #4 as the proper wire size, and this will be the size UF cable to use for the feeder circuit to the barn.

We have found that we need AWG #4 UF cable to keep the voltage drop to within three percent, but in checking our catalogs or inquiring at electrical supply houses we find that the largest size UF cable is AWG #6. We will therefore have to use single wires in the run in place of cable.

Figure 13-3 shows the buildings of Fig. 13-2 with the completed system. It was installed by first removing the panel cover, then removing one of the stamped 1-inch knockouts in the top of the panel box, being careful not to come in contact with any live electrical parts within the box. Insert a 1-inch EMT connector in this knockout, and tighten the locknut to hold it securely.

Next, drill a hole through the sill plate in line with the 1-inch knockout. You could have gone through the masonry wall, but drilling through this wall with a star drill would have taken much more time, plus it would provide an opening to allow moisture to come into the conduit and also in the basement. At this opening in the sill plate you will need a conduit

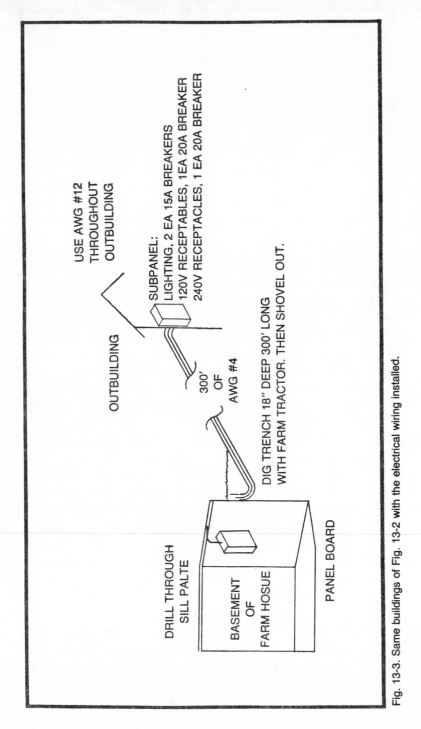

Fig. 13-3. Same buildings of Fig. 13-2 with the electrical wiring installed.

fitting similar to the one in Fig. 13-4 to make the 90-degree bend, and also one outside of the house to make the 90-degree bend to run the conduit below grade.

EMT connectors will be required at these fittings to provide a secure connection of the EMT as you run it from panel connector to the first conduit fitting inside of the basement, then through the sill plate to the other conduit fitting outside of the house, and finally from this fitting to approximately 18 inches below grade. Notice that a 90-degree long-sweep type is on the end underground as well as a fiber bushing to keep the insulation on the wires from becoming damaged.

We will assume that you dug a small hole in the ground near the basement wall to get the conduit 18 inches below grade. But to dig the 300-foot trench by hand would be quite a task, although it could be done. A better way would be to attach a cultivator-type single plow to your tractor and plow open the trench from the house to the building. Then use your shovel to clean out the trench for the wires.

Your next logical step would be to install another run of conduit at the barn to a small subpanel located on the inside of the barn. Install another 90-degree bend in the trench like you did at the house, run the conduit up to the height of the subpanel and use a conduit fitting to make the 90-degree bend at this point. Then run a short nipple from this fitting through the barn wall directly to the back of the subpanel. Your conduit system is now complete. Next comes the wires.

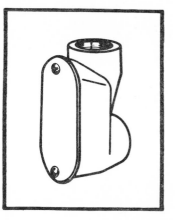

Fig. 13-4. Conduit fitting used to make 90-degree bend and pipe run.

If your soil contains lots of small sharp rocks it will be necessary to shovel in about a 2-inch layer of sand in the bottom of the trench before installing the wires.

Now pull the three current-carrying conductors and one grounding conductor from the panel in the basement to the subpanel in the barn. The wires at the basement panel should be protected with a two-pole, 30-ampere circuit breaker, tieing the two hot wires to it, and connecting the white or neutral wire to the solid neutral bus bar inside of the panelboard. The two hot wires on the other end will connect to the main lugs on the bars, and the neutral again to the neutral bus. The grounding wire will connect at both ends to the isolated grounding terminals inside of each panel.

Before filling the trench another 2 inches of sand should be placed on top of the wires to protect them from sharp stones. It wouldn't hurt to also place a continuous warning ribbon along the trench a few inches below grade in case any digging in the area at a later data is anticipated. Figure 13-5 shows a cross section of a trench with buried wires. Note the sand, warning ribbon, etc.

In the subpanel in the barn you will need two 15-ampere one-pole circuit breakers to protect two lighting circuits, one 15-ampere one-pole circuit breaker for the receptacle circuit and one 20-ampere two-pole circut breaker for the 20-ampere receptacle. From these circuit breakers in the subpanel, you may use AWG #12 for all the circuits, which may be installed and connected as described earlier in this book for similar circuits.

## INSTALLING OUTSIDE RESIDENTIAL LIGHTING

The first area of your home to be considered is the entrance to your home. With proper entrance lighting a visitor's first impression will be a good one. And, at the same time, you'll show your family and friends you really care enough to clear a path through the dark to see them safely to your door.

A post-mounted lighting fixture at the walk is also a good idea. Figure 13-6 shows one way of installing one. The circuit feeding the post lamp begins at the panelboard and runs to a junction box on the outside wall. This wiring can be installed

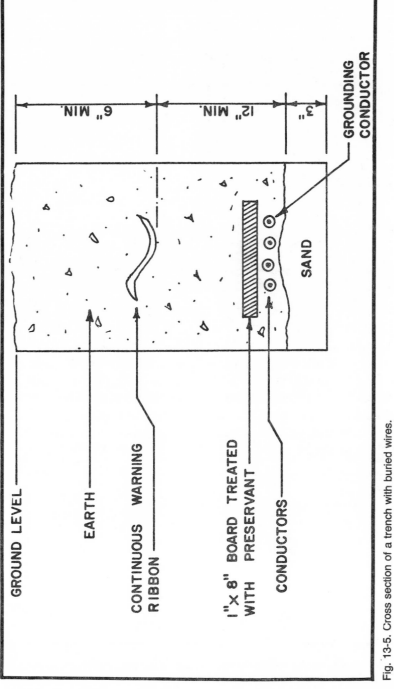

GROUND LEVEL

EARTH

CONTINUOUS WARNING RIBBON

1" × 8" BOARD TREATED WITH PRESERVANT

CONDUCTORS

6" MIN.

12" MIN.

3"

GROUNDING CONDUCTOR

SAND

Fig. 13-5. Cross section of a trench with buried wires.

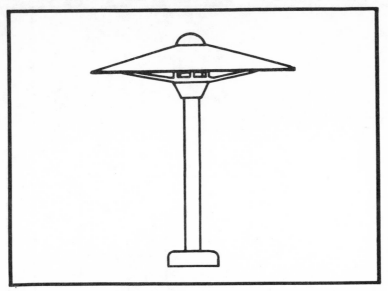

Fig. 13-6. Post-mounted lighting fixture.

by any wiring method approved for indoor wiring, such as NM cable. A piece of conduit is then used from the junction box through the mansonry wall to a foot or so beyond the wall on the outside of the house, and at least 18 inches below grade. Both ends of this conduit have fiber bushings to protect the wire from cuts and other damage. A splice is made within the junction box to connect the NM cable to direct-burial-type UF cable. The cable then runs from the junction box to the connection point of the post lamp. In this case, the wires are

Fig. 13-7. Outdoor lighting enables families to participate in outdoor activities after dark.

184

Fig. 3-8. Dusk-to-dawn light used to illuminate entrance roadways to homes and farms.

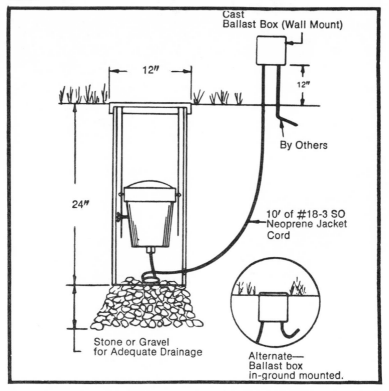

Cast
Ballast Box (Wall Mount)

12"

12"

By Others

24"

10' of #18-3 SO
Neoprene Jacket
Cord

Stone or Gravel
for Adequate Drainage

Alternate—
Ballast box
in-ground mounted.

Fig. 13-9. Decorative lighting, like this well light, can be used to great advantage in improving the outside appearance of one's house or lawn during night hours.

Fig. 13-10. Outdoor lighting installed around farm yards enables farmers to perform tasks after dark safer and with greater efficiency.

enclosed in a sheathed cable and consists of two AWG #12 wires and one grounding wire.

Outdoor patios may be lighted by a combination of wall-mounted fixtures on the side of your home adjacent to the patio, floodlights mounted under the eaves of your house or low-level mushroom lighting fixtures installed along the

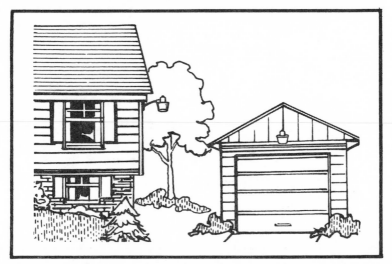

Fig. 13-11. Outdoor lighting around the home tends to deter prowlers while at the same time invites guests.

186

perimeter of the patio. When selecting these fixtures from manufacturer's catalogs always choose types that match the architectural character of the home.

Lighting colorful shrubs and flowers is another method of decorating the outside of your home. Try installing ground-mounted up-lights under trees on the lawn that will shine up through the tree branches. This will add elegance to any lawn and home.

Figures 13-7 through 13-11 show several outdoor lighting schemes employed by homeowners to enhance the appearance of their home and to extend the outside activities beyond darkness. You will be able to get other ideas from residential lighting catalogs. Addresses of some are listed in Appendix III.

Personal taste is the final factor to be considered in the process of selecting outdoor lighting equipment, but always try to get ones that are durable as outdoor fixtures. Normally these have to stand a great amount of abuse throughout the year.

# Chapter 14
# What About Motors?

In every household and around every farm there is a need for electric motors...many electric motors. Your refrigerator uses a hermetically sealed motor to help keep your food cool. When your forced air system calls for heat or cooling a motor drives the blower fan to force the air through the ducts and out the air outlets. The list of appliances and equipment powered by electric motors does not end here. Vacuum cleaners, water pumps, conveyor belts, ventilating fans, sewing machines, and many other similar pieces of equipment depend upon electric motors to make the equipment function.

When you purchase an appliance or piece of equipment with built-in motors the motor has already been sized by qualified engineers and the proper type selected for the best operation to run the appliance. However, there will be times when you will have to size and select your own motor for custom appliances often needed in the home or for use around the farm. For example, you might need a bench grinder or buffing wheel, an air compressor or a paint sprayer. Or maybe you have constructed a fan to provide ventilation for one of your stock-holding buildings. What size and type of motor will you use to drive these various items?

The selection of the right motor to do the job involves several factors, but the following will serve as a basic guide in doing so.

- The motor must be of the correct physical size to fit the piece of equipment in which it is to be installed
- It must be rated at the correct horsepower to drive the load which the equipment is intended to accomplish
- The motor must be wound for the correct voltage characteristics, that is, the proper line voltage and phase to match that of the electrical system to which it will be connected
- Speed is an important factor if the motor is connected directly to the load. However, you will have some degree of margin when the motor is connected to gear or belt drives as the desired speed can be obtained by adjusting the pulley or gear ratio when coupling the motor to its load
- The type of the motor is also important. For the same horsepower there are many different types of motors for different types of load, and you must select the proper type for the best results

To give you the basis for selecting the proper size and type of motor for a given application you should first have a knowledge of the operating characteristics of electric motors.

Most motors used in the home and around the farm are of the single-phase type for use on either 120- or 240-volt lines. They are further divided into types that differ from one another by the methods used in starting them; the main types being:

- Capacitor-start
- Split-phase
- Capacitor-start, capacitor-run
- Repulsion-start, induction-run
- Universal
- Shaded-pole

It would take several volumes to describe in detail the operating and starting characteristics of each of these types, so we will give only a brief description of each and the most common loads for which the type is used.

## CAPACITOR-START MOTOR

This type of motor is wound and wired so that a capacitor is connected in series with the starting winding to give this type of motor a high starting torque for use on appliances that are likely to be difficult to start. Capacitor-start motors are normally manufactured in sizes from ⅛ to 7½ horsepower and are well suited for use on refrigerators, washers, pumps, milking machines and similar types of appliances.

## SPLIT-PHASE MOTOR

Motors of the split-phase type have a special starting winding that limits the starting torque to approximately twice that of the full-load torque. Once the motor has reached a certain given speed, a centrifugal switch cuts out the starting winding and the motor runs like a regular induction motor. This type of motor is made in sizes from 1/20 to 1/3 horsepower and is best suited for use on loads where the full load will not be applied until after the device or appliance reaches its full speed. One example of its use would be to drive a bench grinder. The motor is started under a very light load to start the grinding wheel revolving. When the motor reaches full speed the grinder can be used to sharpen bits and similar items, which makes the motor work at its full capacity since pressure is placed against the wheel. Other uses would include small ventilating fans and small stationary shop tools such as drill presses, buffing machines, and sanders.

## CAPACITOR-START CAPACITOR-RUN MOTOR

This type of motor starts in the same way that a regular capacitor-start motor does, and, in addition, has a capacitor in the running circuit as well as the starting circuit. However, the starting-winding circuit cuts out when the motor reaches its full speed. This type of motor has high operating efficiency and is well suited for use on appliances where the load is connected directly to the motor without belts, pulleys or gears. Such applications include refrigerators, small centrifugal pumps, and ventilating fans requiring a drive rated from ½ to 10 horsepower.

## REPULSION-START INDUCTION-RUN MOTOR

This type of motor is started by means of two magnetic fields that causes the rotor to turn. Once the rotor reaches its running speed the brushes are automatically lifted away from the rotor and the motor then runs as an induction motor. Motors of this type are well suited for driving equipment that have heavy starting loads such as chain conveyors, reciprocating pumps, air compressors, feed grinders, etc. They are available in sizes from ½ to 10 horsepower.

## UNIVERSAL MOTOR

A universal motor, as the name implies, will operate on either DC or AC circuits, and is used to drive electric hand tools, vacuum cleaners, food mixers, sewing machines, and similar household appliances. This type of motor should always be operated under load as this is the only means of speed control. If operated under no load, this type of motor will reach very high and dangerous speeds. Most of these motors are manufactured in fractional horsepower only.

## SHADED-POLE MOTOR

In place of auxiliary windings in split-phase motors, the shaded-pole motor has a squirrel-cage rotor and copper loops across the stator coils to produce the shaded-pole effect. It is always made in fractional horsepower only and can start only very light loads. Such loads might include electric clocks, a small bathroom fan, a film projector, etc. Its restricted use is offset by the low cost of such motors.

## ELECTRIC MOTOR MAINTENANCE

The proper care of electric motors is very important to insure that each will have a long life. But few motors used around the home or farm ever get a fraction of the care they need. Most homeowners purchase a new appliance, run it hard for as long as it will last, then purchase a new one. This is sad indeed because with a little care electric motors can outlive the appliance that they are driving. To verify this fact take a look at the many people who are using appliances over 30 years old. Also, when these same people do buy new

191

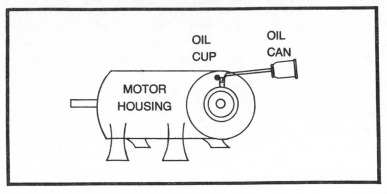

Fig. 14-1. Method of oiling motor bearings.

appliances to replace the old, the electric motor will still be good nine out of ten times.

Here are a few items that will prolong the life of your electric motors and will also enable them to operate at their highest efficiency.

Make sure that you use the proper wire size for the electric circuit feeding the motor to keep voltage drop to the minimum. Low voltage not only lowers the efficiency of the motor but also shortens its life considerably.

Keep the coils of an open or dripproof motor clean by blowing the dust out of the coils at regular intervals. The frequency of these intervals will depend on the kind of environment the motor is used. Dirty coils will cause the motor to overheat from lack of internal ventilation and will cause the insulation to break down sooner than it should.

Put a few drops of oil in the bearings of motors that require lubrication as shown in Fig. 14-1. But don't use too much, since excess might get into internal switch contacts or else the inside of the motor and damage the insulation of the coils. Always use the type of oil recommended by the manufacturer.

Keep the belt pulley⌐ and the shafts on which they are mounted properly aligned, as other the belt will soon wear out. The pulleys should be aligned so that a straightedge placed across their faces (Fig. 14-2) contacts both edges of both pulleys. Figure 14-3 shows two examples of pulleys that are improperly aligned.

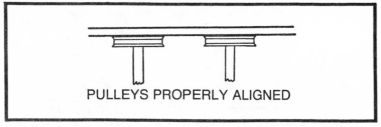

PULLEYS PROPERLY ALIGNED

Fig. 14-2. Example of properly aligned motor pulleys.

Also make certain that all pulley belts are under the proper tension. Too loose a belt causes slippage and excessive belt and pulley wear while a too tight a belt will cause excessive wear on the motor bearings. A rule of thumb for adjusting pulley belts for proper tension is to tighten the belts so that approximately a half-inch can be depressed between motor and the machine being driven as shown in Fig. 14-4.

Unless specifically designed for wet or damp locations, keep all motors free from excessive moisture as this will .cause the winding insulation to break down much faster than it should.

Keep motor mounts tight to prevent excessive vibration and to keep the pulleys and shafts aligned. Either will cause damage to both the belts and the motor bearings.

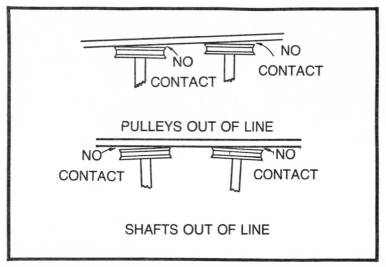

Fig. 14-3. Examples of improperly aligned motor pulleys.

193

## Table 14-1. Motor Overcurrent Protection Ratings

| MOTOR AMPERAGE | MOTOR HORSEPOWER 1-PHASE 115V | 1-PHASE 230V | 3-PHASE 208V | 3-PHASE 230V | 3-PHASE 460V | 3-PHASE 575V | 2-PHASE 230V | 2-PHASE 460V | FUSE AMPERE RATING STD[1] | HD[2] | MAX SIZE 40°C | MAX SIZE ALL OTHERS |
|---|---|---|---|---|---|---|---|---|---|---|---|---|
| 1.01 to 1.04 | | | | | ½ | | | ½ | 1 | 1⅛ | 1 4/10 | 1¼ |
| 1.05 to 1.08 | | | | | | | | | 1⅛ | 1¼ | 1 4/10 | 1¼ |
| 1.09 to 1.14 | | | | | | ¾ | | | 1⅛ | 1¼ | 1 4/10 | 1 4/10 |
| 1.15 to 1.17 | | | | | | | | | 1⅛ | 1¼ | 1 6/10 | 1 4/10 |
| 1.18 to 1.21 | | | | | | | | ¾ | 1¼ | 1 4/10 | 1 6/10 | 1 4/10 |
| 1.22 to 1.28 | | | | | | | | | 1¼ | 1 4/10 | 1 6/10 | 1 6/10 |
| 1.29 to 1.30 | | | | | | | | | 1¼ | 1 4/10 | 1 8/10 | 1 6/10 |
| 1.31 to 1.39 | | | | | | | | | 1 4/10 | 1 6/10 | 1 8/10 | 1 6/10 |
| 1.40 to 1.42 | | | | | ¾ | 1 | | | 1 4/10 | 1 6/10 | 1 8/10 | 1 8/10 |
| 1.43 to 1.46 | | | | | | | | | 1 4/10 | 1 6/10 | 2 | 1 8/10 |
| 1.47 to 1.56 | | | | | | | | | 1 6/10 | 1 8/10 | 2 | 1 8/10 |
| 1.57 to 1.60 | | | | | | | | 1 | 1 6/10 | 1 8/10 | 2 | 2 |
| 1.61 to 1.66 | | | | | | | | | 1 6/10 | 1 8/10 | 2¼ | 2 |
| 1.67 to 1.73 | | | | | | | | | 1 8/10 | 2 | 2¼ | 2 |
| 1.74 to 1.80 | | | | | 1 | | | | 1 8/10 | 2 | 2¼ | 2¼ |
| 1.81 to 1.87 | | | | | | | | | 1 8/10 | 2 | 2½ | 2¼ |
| 1.88 to 1.95 | | | | | | | | | 2 | 2¼ | 2½ | 2¼ |
| 1.96 to 2.00 | | | | ½ | | | ½ | | 2 | 2¼ | 2½ | 2½ |
| 2.01 to 2.08 | | | | | | | | | 2 | 2¼ | 2 8/10 | 2½ |
| 2.09 to 2.17 | | | | | | 1½ | | | 2¼ | 2½ | 2 8/10 | 2½ |
| 2.18 to 2.28 | | 1/6 | ½ | | | | | | 2¼ | 2½ | 2 8/10 | 2 8/10 |
| 2.29 to 2.34 | | | | | | | | 1½ | 2¼ | 2½ | 3 2/10 | 2 8/10 |
| 2.35 to 2.43 | | | | | | | ¾ | | 2½ | 2 8/10 | 3 2/10 | 2 8/10 |
| 2.44 to 2.56 | | | | | | | | | 2½ | 2 8/10 | 3 2/10 | 3 2/10 |
| 2.57 to 2.60 | | | | | 1½ | | | | 2½ | 2 8/10 | 3½ | 3 2/10 |
| 2.61 to 2.78 | | | | | | 2 | | | 2 8/10 | 3 2/10 | 3½ | 3 2/10 |
| 2.79 to 2.85 | | | | ¾ | | | | | 2 8/10 | 3 2/10 | 3½ | 3½ |
| 2.86 to 2.91 | | ¼ | | | | | | | 2 8/10 | 3 2/10 | 4 | 3½ |
| 2.92 to 3.04 | | | | | | | | 2 | 3 2/10 | 3½ | 4 | 3½ |
| 3.05 to 3.21 | | | ¾ | | | | 1 | | 3 2/10 | 3½ | 4 | 4 |
| 3.22 to 3.33 | | | | | | | | | 3 2/10 | 3½ | 4½ | 4 |
| 3.34 to 3.47 | | | | | 2 | | | | 3½ | 4 | 4½ | 4 |
| 3.48 to 3.60 | | 1/3 | | 1 | | | | | 3½ | 4 | 4½ | 4½ |
| 3.61 to 3.64 | | | | | | | | | 3½ | 4 | 5 | 4½ |
| 3.65 to 3.91 | | | | | | 3 | | | 4 | 4½ | 5 | 4½ |
| 3.92 to 4.00 | | | 1 | | | | | | 4 | 4½ | 5 | 5 |
| 4.01 to 4.16 | | | | | | | | | 4 | 4½ | 5 6/10 | 5 |
| 4.17 to 4.34 | | | | | | | | 3 | 4½ | 5 | 5 6/10 | 5 |
| 4.35 to 4.48 | 1/6 | | | | | | | | 4½ | 5 | 5 6/10 | 5 6/10 |
| 4.49 to 4.68 | | | | | | | 1½ | | 4½ | 5 | 6¼ | 5 6/10 |
| 4.69 to 4.86 | | | | | 3 | | | | 5 | 5 6/10 | 6¼ | 5 6/10 |
| 4.87 to 5.00 | | ½ | | | | | | | 5 | 5 6/10 | 6¼ | 6¼ |
| 5.01 to 5.20 | | | | 1½ | | | | | 5 | 5 6/10 | 7 | 6¼ |
| 5.21 to 5.43 | | | | | | | | | 5 6/10 | 6¼ | 7 | 6¼ |
| 5.44 to 5.71 | | | 1½ | | | | | | 5 6/10 | 6¼ | 7 | 7 |
| 5.72 to 5.82 | ¼ | | | | | | | | 5 6/10 | 6¼ | 8 | 7 |
| 5.83 to 6.08 | | | | | | | 2 | | 6¼ | 7 | 8 | 7 |
| 6.09 to 6.42 | | | | | | 5 | | | 6¼ | 7 | 8 | 8 |
| 6.43 to 6.50 | | | | | | | | | 6¼ | 7 | 9 | 8 |
| 6.51 to 6.95 | | ¾ | | 2 | | | | 5 | 7 | 8 | 9 | 8 |
| 6.96 to 7.20 | 1/3 | | | | | | | | 7 | 8 | 9 | 9 |
| 7.21 to 7.28 | | | | | | | | | 7 | 8 | 10 | 9 |
| 7.29 to 7.82 | | | 2 | | 5 | | | | 8 | 9 | 10 | 9 |
| 7.83 to 8.32 | | 1 | | | | | 3 | | 8 | 9 | 10 | 10 |
| 8.33 to 8.57 | | | | | | | | | 9 | 10 | 10 | 10 |
| 8.58 to 8.69 | | | | | | | | | 9 | 10 | 12 | 10 |
| 8.70 to 9.36 | | | | | | 7½ | | 7½ | 9 | 10 | 12 | 12 |
| 9.37 to 10.4 | ½ | 1½ | | 3 | | | | | 10 | 12 | 12 | 12 |
| 10.5 to 10.7 | | | | 3 | | | | | 12 | 12 | 12 | 12 |
| 10.8 to 12.4 | | 2 | | | 7½ | 10 | | 10 | 12 | 15 | 15 | 15 |
| 12.5 to 13.0 | | | | | | | | | 15 | 15 | 17½ | 15 |
| 13.1 to 14.2 | ¾ | | | | 10 | | 5 | | 15 | 17½ | 17½ | 17½ |

1—Standard service
2—Heavy duty service
3—Applies to motors marked to have a temperature rise not over 40°C or marked with a service factor not less than 1.15. shown in next column
All other motors take maximum

# Table 14-1 (continued from page 194)

| MOTOR AMPERAGE | MOTOR HORSEPOWER | | | | | | | | FUSE AMPERE RATING | | | |
|---|---|---|---|---|---|---|---|---|---|---|---|---|
| | 1-PHASE 115V | 1-PHASE 230V | 3-PHASE 208V | 3-PHASE 230V | 3-PHASE 460V | 3-PHASE 575V | 2-PHASE 230V | 2-PHASE 460V | STD[1] | HD[2] | MAX SIZE 40°C | MAX SIZE ALL OTHERS |
| 14.3 to 15.2 | | | | 5 | | | | | 15 | 17½ | 20 | 17½ |
| 15.3 to 15.6 | | | | | | | | | 15 | 17½ | 20 | 20 |
| 15.7 to 17.8 | 1 | 3 | 5 | | | 15 | | | 17½ | 20 | 20 | 20 |
| 17.9 to 20.8 | 1½ | | | | | | 7½ | 15 | 20 | 25 | 25 | 25 |
| 20.9 to 21.4 | | | | | 15 | | | | 25 | 25 | 25 | 25 |
| 21.5 to 21.7 | | | | | | | | | 25 | 25 | 30 | 25 |
| 21.8 to 24.9 | 2 | | 7½ | 7½ | | 20 | 10 | 20 | 25 | 30 | 30 | 30 |
| 25.0 to 26.0 | | | | | | | | | 25 | 30 | 35 | 30 |
| 26.1 to 28.5 | | 5 | | 10 | 20 | 25 | | | 30 | 35 | 35 | 35 |
| 28.6 to 30.4 | | | | | | | | 25 | 30 | 35 | 40 | 35 |
| 30.5 to 32.1 | | | 10 | | | 30 | | | 35 | 40 | 40 | 40 |
| 32.2 to 34.7 | 3 | | | | 25 | | | | 35 | 40 | 45 | 40 |
| 34.8 to 36.0 | | | | | | | 15 | 30 | 35 | 40 | 45 | 45 |
| 36.1 to 39.1 | | | | | | | | | 40 | 45 | 50 | 45 |
| 39.2 to 41.2 | | 7½ | | | 30 | 40 | | | 40 | 45 | 50 | 50 |
| 41.3 to 42.8 | | | | 15 | | | | | 45 | 50 | 50 | 50 |
| 42.9 to 43.4 | | | | | | | | | 45 | 50 | 60 | 50 |
| 43.5 to 46.0 | | | | | | | | 40 | 45 | 50 | 60 | 60 |
| 46.1 to 49.9 | | | 15 | | | | 20 | | 50 | 60 | 60 | 60 |
| 50.0 to 51.0 | | 10 | | | | | | | 50 | 60 | 70 | 60 |
| 51.1 to 52.1 | | | | | 40 | 50 | | | 60 | 60 | 70 | 60 |
| 52.2 to 57.1 | 5 | | | 20 | | | | 50 | 60 | 70 | 70 | 70 |
| 57.2 to 60.0 | | | 20 | | | | 25 | | 60 | 70 | 70 | 70 |
| 60.1 to 64.2 | | | | | | 60 | | | 60 | 80 | 80 | 80 |
| 64.3 to 69.5 | | | | 25 | 50 | | 30 | 60 | 70 | 80 | 90 | 80 |
| 69.6 to 72.0 | | | | | | | | | 70 | 80 | 90 | 90 |
| 72.1 to 78.2 | | | 25 | | 60 | 75 | | | 80 | 90 | 100 | 90 |
| 78.3 to 78.5 | | | | | | | | | 80 | 90 | 100 | 100 |
| 78.6 to 82.0 | 7½ | | | 30 | | | | | 80 | 90 | 110 | 100 |
| 82.1 to 86.9 | | | | | | | | 75 | 90 | 100 | 110 | 100 |
| 87.0 to 89.2 | | | 30 | | | | | | 90 | 100 | 110 | 110 |
| 89.3 to 91.5 | | | | | | | 40 | | 90 | 100 | 125 | 110 |
| 91.6 to 95.6 | | | | | | | | | 100 | 110 | 125 | 110 |
| 95.7 to 100 | 10 | | | | 75 | 100 | | | 100 | 110 | 125 | 125 |
| 101 to 108 | | | | 40 | | | | | 110 | 125 | 125 | 125 |
| 109 to 114 | | | 40 | | | | 50 | 100 | 110 | 125 | 150 | 150 |
| 115 to 124 | | | | | 100 | | | | 125 | 150 | 150 | 150 |
| 125 to 130 | | | | 50 | | 125 | | | 125 | 150 | 175 | 150 |
| 131 to 142 | | | | | | | 60 | 125 | 150 | 175 | 175 | 175 |
| 143 to 152 | | | 50 | | | 150 | | | 150 | 175 | 200 | 175 |
| 153 to 160 | | | | 60 | 125 | | | 150 | 175 | 200 | 200 | 200 |
| 161 to 173 | | | 60 | | | | 75 | | 175 | 200 | 225 | 200 |
| 174 to 177 | | | | | | | | | 175 | 200 | 225 | 225 |
| 178 to 180 | | | | | 150 | | | | 200 | 225 | 225 | 225 |
| 181 to 195 | | | | 75 | | 200 | | | 200 | 225 | 250 | 225 |
| 196 to 200 | | | | | | | | | 200 | 225 | 250 | 250 |
| 201 to 214 | | | 75 | | | | | 200 | 225 | 250 | 250 | 250 |
| 215 to 217 | | | | | | | | | 225 | 250 | 300 | 250 |
| 218 to 230 | | | | | | | 100 | | 225 | 250 | 300 | 300 |
| 231 to 249 | | | | 100 | 200 | | | | 250 | 300 | 300 | 300 |
| 250 to 260 | | | | | | | | | 250 | 300 | 350 | 300 |
| 261 to 285 | | | 100 | | | | 125 | | 300 | 350 | 350 | 350 |
| 286 to 304 | | | | | | | | | 300 | 350 | 400 | 350 |
| 305 to 321 | | | | 125 | | | | 150 | 350 | 400 | 400 | 400 |
| 322 to 347 | | | 125 | | | | | | 350 | 400 | 450 | 400 |
| 348 to 350 | | | | | | | | | 350 | 400 | 450 | 450 |
| 351 to 360 | | | | 150 | | | | | 400 | 450 | 450 | 450 |
| 361 to 391 | | | | | | | | | 400 | 450 | 500 | 450 |
| 392 to 400 | | | 150 | | | | | | 400 | 450 | 500 | 500 |
| 401 to 428 | | | | | | | | 200 | 450 | 500 | 500 | 500 |
| 429 to 434 | | | | | | | | | 450 | 500 | 600 | 500 |
| 435 to 450 | | | | | | | | | 450 | 500 | 600 | 600 |

†This applies to motors marked to have a temperature rise not over 40° C. or marked with a service factor of not less than 1.15. All other motors take maximum shown in fourth column. (430-32a1)

If two ampere ratings are shown on motor, make sure for what voltage motor is connected. Horse power of motor gives no definite indication of size Fusetron or Low-Peak fuse to be used as amperage of AC motors of the same size varies a great deal.

Note: Use Fusetron fuses on D.C. motors. See page 20.

‡In ordering Fusetron D.E. Fuses Specify FRN plus Amps. for 250V.; FRS plus Amps. for 600V. In ordering Low-Peak D.E. Fuses Specify LPN plus Amps. for 250V.; LPS plus Amps. for 600V.

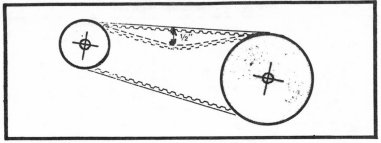

Fig. 14-4. Example of proper belt tension between pulleys.

The appliance or machine that is being driven is often overlooked during maintenance of electric motors, but this is just as important. A faulty bearing in the shaft of the machine being driven can cause an excessive overload and damage the motor. However, if the motor circuit is provided with the proper overload protection the circuit will usually open before any harmful damage is done to the motor under such conditions. Therefore, always make certain that each of your motors are fused with the recommended fuse or circuit breaker size (see Table 14-1) and also provided with an overload relay or "heater" as they are sometimes called.

## TYPICAL APPLICATIONS OF ELECTRIC MOTORS

The bench grinder is a simple tool needed around every farm and home shop for sharpening metal drill bits, axes, knives and similar tools, as well as buffing various items and removing burrs from metal objects. A quarter-horsepower motor will be just about right and, as mentioned previously, the motor does not have to have a high starting torque since grinders are started with little or no load.

If your motor has a speed of around 3600 RPM you can connect the grinding wheel directly to the motor shaft by means of a simple motor arbor designed for the size wheel you intend to use. But if you will be using a slower speed motor, such as the common 1800 RPM ones, you will need a V-belt drive having a pulley ratio of about two to one to produce the correct speed at the grinding wheel. A split-phase motor would be ideal for this application.

Assume that you have just purchased or constructed a compact power saw with tilting or rolling table to saw logs up to 12 inches thick. Now you must select and install an electric motor for this saw. Always check with the manufacturer's recommendations, but the majority of these table saws will operate best with a 1-horsepower repulsion-start induction-run type motor rated at 1800 RPM. Both the motor and saw frame should be firmly secured and aligned to accept a pulley V-belt drive.

If your motor speed does not correspond to the speed required by the machine, usually the speed can be changed on one or the other by means of pulley wheels or gears. For determining pulley sizes and speeds, the following formula may be used,

$$\text{Machine pulley dia} = \frac{\text{motor pulley dia} \times \text{motor RPM}}{\text{machine RPM}}$$

In using this formula all pulley diameters should be calculated in inches. For example, we have just installed a motor rated at 1720 RPM with an 8-inch motor pulley to drive a feed mill at 3600 RPM. What size pulley should be used on the mill to obtain this speed? Substitute known values in the formula to get,

$$\text{Machine pulley dia} = \frac{8(\text{inches}) \times 1720\,(\text{RPM})}{3600\,(\text{RPM})} = 3.82$$

Therefore, a 4-inch pulley on the mill would give the approximate required RPM.

The following table gives some of the most common appliances used in the household or around the farm, the size motor (in hp) most used on each type, and the recommended motor type to use in case a replacement is needed or else you acquired the machine without a motor.

### Application of Fractional Horsepower Motors

| MACHINE | NORMAL HP | RECOMMENDED MOTOR TYPE |
|---|---|---|
| Churn | 1/6 or 1/4 | split-phase |
| Concrete mixer | 1/4 to 1/2 | repulsion induction |
| Corn sheller | 1/4 | capacitor-start |
| Cream separator | 1/4 | capacitor-start |
| Fanning mill | 1/4 | split-phase |
| Farm/home shop equip | 1/4 to 1/2 | capacitor-start/ repulsion induction |

| MACHINE | NORMAL HP | RECOMMENDED MOTOR TYPE |
| --- | --- | --- |
| Fruit grader | ¼ | repulsion induction |
| Bench grinder | ¼ | split-phase |
| Potato grader | ½ | repulsion induction/capacitor-start |
| Hydraulic pump | ½ | repulsion induction/capacitor-start |
| Root cutter | ½ to 1 | capacitor-start |
| Meat grinder | ¼ | split-phase |
| Sheering tool | ¼ | capacitor-start |
| Small feed grinder | ½ to 1 | repulsion induction/capacitor-start |
| | ¼ | |
| Washing machine | | capacitor-start |

## MOTOR REPAIRS

If the winding in a fractional horsepower motor burns up it is usually less expensive to replace it with a new one than it would be to attempt a repair. However, there are some faults that occur on these motors that the homeowner or farmer can easily repair. Here are some of the most common troubles and the method of repairing them; most will occur on the universal type motors. When a motor will not run, check the following in the order given:

1. No current at the outlet; Check with a voltmeter. If no current exists determine the trouble and replace fuse or reset circuit breaker. The reason for the blown fuse could be a shorted cord.

2. A damaged attachment plug or open or shorted cord. Check the cord with an ohmmeter as described earlier. Repair or replace.

3. A broken switch on the appliance. Open the wiring to switch and test with an ohmmeter. Replace if defective.

4. Broken, shattered or worn brushes. Open brush holder and inspect. If broken or worn replace with new ones.

5. Motor windings burned out. Check each with ohmmeter or better yet, with your nose. Usually you can immediately tell by the odor if the windings are burned out. If so, replace with a new motor on smaller ones (under 1 horsepower) or have the larger sizes rewound at a motor repair shop.

If none of these five faults seem to be causing the trouble, it will be best to call in a repairman to find the trouble.

# Chapter 15
# Why Overcurrent Protection?

Use of electricity, when its systems are properly installed and provided with overcurrent protection, is one of the safest and most efficient forms of energy known to man. On the other hand improperly installed electrical systems can not only cause hazards to property in the form of fires, but can also cause harmful injuries to people in the form of deadly electrical shock.

Therefore, every wiring system, no matter what type it may be, must be installed properly using approved materials for the purpose and in a workmanlike manner using methods approved by the National Electrical Code. One strict requirement of the National Electrical Code is that every electrical circuit be provided with proper overcurrent protection. Note the word *proper*.

Sometimes, when a fuse blows, some person who doesn't understand the function and safety value of overcurrent protective devices may replace the blown fuse with one of a higher rating (a 30-ampere fuse in place of the original 15-ampere fuse), a piece of copper wire, or in some cases, even places a copper penny behind the blown fuse in order to allow the current to pass through the penny with no protection. All are exceedingly dangerous practices and should never be duplicated under any circumstances.

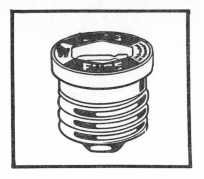

Fig. 15-1. Plug fuses are made with a threaded base.

Overcurrent devices in electrical systems are similar in purpose to safety valves on steam boilers. With a boiler should the steam pressure rise to a point that it is unsafe, more than the strength of the boiler should stand, the safety valve opens and relieves this pressure. In electrical circuits whenever the current load becomes more than the conductors should safely carry the fuses blow or the circuit breakers trip and opens the circuit, which stops the current flow. This prevents damage to the conductor insulation, the equipment being fed and other components on the circuit. It also prevents the wires from heating to a point where they could cause fires. The circuit could be a high-voltage transmission line or a 120-volt lighting circuit, but each must have some form of protection of the proper size and type.

In residential and farm electrical systems overcurrent protective devices consist of fuses and circuit breakers that are usually factory assembled and placed in a metal cabinet commonly called a load center or panelboard.

Fusible service equipment or fuse blocks are rated at 15, 20, 30, 60, 100, 200, 400 amperes etc. Although the equipment themselves do not have any inbetween ratings, the fuses installed in the equipment may have any rating below the equipment rating. For example, assume that a piece of electrical equipment like an electric range requires an overcurrent protection of 40 amperes. A 60-ampere switch or fuse block will have to be installed, but only 40-ampere fuses will be inserted in the fuse holders. Therefore, the wires feeding this range need only be rated for 40 amperes also.

Fusible panelboards are available containing main fuse blocks to kill (deactivate) the entire panel when pulled. In addition they contain other fuse spaces for the connection of from two to forty branch circuits. One fuse space is needed for each 120-volt circuit, and two spaces are required for each 240-volt circuit.

Fuses are made in many different styles and sizes for different voltages and current loads, but they all operate on the same general principle; that is, when excessive current flows through the fuse a piece of soft metal within the fuse becomes overheated and melts the metal link. This in turn opens the circuit. The temperature at which the fuse link melts depends upon the amount of excess current, the duration of excess current and the ease with which heat escapes from the fuse. If the fuse did not open the circuit when excessive current occurred for any length of time the wiring would burn, equipment would be damaged or a fire or personal injury would occur.

So by now we should readily see the great importance of having overcurrent protection of the proper size and type in every electrical circuit.

## PLUG FUSES

Plug fuses are made with ratings from 1 to 30 amperes. These fuses are the type most commonly used for overcurrent protection in residential wiring systems. All are made with a threaded base (see Fig. 15-1) which screws into a socket similar to a lamp socket. Several types of plug fuses are shown in Fig. 15-2.

Up until 1935 all plug fuses were made with the same threaded base known as the Edison base. However, electrical inspectors found that many people, ignorant of the importance

Fig. 15-2. Illustration showing different types of plug fuses.

of proper overcurrent protection, would use fuses of too large a size or else used pennies or other materials to bridge the fuse. This, of course, wiped out all protection.

In order to guard against improper use of plug fuses a different base was developed to prevent the use of pennies or other bridging material, making it virtually impossible to tamper with the fuse protection. This type of fuse is made to fit into regular Edison base fuseholders by means of a simple inexpensive adpater that locks in place. The altered fuse base can then be removed or inserted in the same manner as an ordinary fuse.

## CARTRIDGE FUSES

If the circuits to be protected require a rating of over 30 amperes, cartridge fuses are necessary. Figure 15-3 shows two types of cartridge fuses which are commonly used in the home and around the farm. The forms of these two fuses were adopted in 1904 and the dimensions were standardized and are still in use today. The standard cartridge fuse in Fig. 15-3A is rated from 1 to 60 amperes. Fuses are held in place by spring clips on the fuse block which grip the metal ferrule at the end of the cartridge. This makes them very easy and quick to renew when one blows out. The metal fuse link is enclosed inside of the fiber cartridge which prevents its temperature from being affected by air currents. Also the fiber cartridge keeps the molten metal confined on the inside of the cartridge when the fuse blows.

Large sizes of fuses from 65 to 600 amperes are made in the knife-blade type. These as the name implies have a short blade attached to each end caps as shown in Fig. 15-3B.

Another type of fuse that is very useful, especially on motor circuits, is the time-lag fuse. This type has a fuse link element and a thermal cutout element. On overloads the circuit is opened by the thermal cutout. This thermal cutout has a very long time-lag so that the fuse will not open on harmless overloads or ordinary motor-starting currents.

The fuse link is made heavier than those used in an ordinary fuse of the same rating. It protects only against short circuit. This type of fuse will open on short circuits as safely as an ordinary fuse.

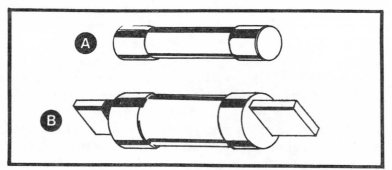

Fig. 15-3. Larger rated fuses. (A) Standard cartridge fuse manufactured in ratings from 1 to 60 amperes. (B) Knife-blade type fuses are rated from 65 to 600 amperes.

This type of fuse construction has the added advantage of a lesser electrical resistance than ordinary fuses so that it will not cause as much heating as ordinary fuses.

In selecting any fuse or fuse holder for use in the home or farm electrical system they should be approved and labeled by Underwriters' Laboratory, Inc., because some cheap unapproved types have been known to cause injury.

## CIRCUIT BREAKERS

A circuit breaker resembles an ordinary toggle switch, and it is probably the most widely used means of overcurrent protection in the home today. While fuses have advantages over circuit breakers for certain types of industrial applications circuit breakers are the choice for overcurrent protection for most residential and farm electrical systems.

On an overload the circuit breaker opens itself or trips automatically. In a tripped position, the handle jumps to the middle position as shown in Fig. 15-4. To reset turn the handle to the off position, then push it as far as it will go in this same direction; next, turn it to the on position again.

One single-pole breaker is required to protect each 120-volt circuit while one double-pole breaker is required to protect each 240-volt single-phase circuit. If the circuit is protecting a three-phase circuit, then a three-pole breaker will be required. Circuit breakers are rated in amperes, just like regular fuses, although the particular ratings are not exactly the same for circuit breakers as those for fuses.

Circuit-breaker enclosures are manufactered in several types. One type contains only branch-circuit breakers, while another contains a main-circuit breaker to kill all circuits contained in the enclosure as well as additional branch-circuit breakers. Most of the circuit breaker types used for residential and farm applications are of the plug-in type, which means that the cabinets or enclosures are sold in various ratings but without circuit breakers. They have only the bus bars and provisions for attaching the breakers. The user then selects whatever combination of circuit breakers required for the protection of the circuits in his electrical system and plugs them into this bus bar arrangement.

## SELECTING PROPER OVERCURRENT PROTECTION

In general the fuse size should never exceed the rating of the wires themselves. For example, Table III-1 in Appendix III gives the current carrying capacity of most types of wires used in home and farm electrical systems. If the wire size in a given circuit is, say, AWG #14 the maximum fuse size should be rated at 15 amperes. If AWG #12 the maximum fuse size should be 20 amperes. The only exception to this rule is when standard fuses are used to protect motor circuits. Then the fuse size may be increased slightly to compensate for the higher starting current required on motor circuits. Table 14-1 gives the proper fuse sizes for motors of various horsepower; these meet requirements set forth in the National Electrical Code.

For other circuits the size fuse or circuit breaker may be determined by the nameplate rating of the circuit. If an electric heating circuit, for example, has a load of 5600 watts of resistance type heaters, what size wire and overcurrent protection would be required? The applied voltage is 240 volts. First the amperage is calculated by

$$\text{Amperes} = \frac{5600 \text{ (watts)}}{240 \text{ (volts}} \times 1.25$$

$$= 29.16$$

From Table III-1 in Appendix III we see that the rating closest to the calculated load without going under it is 30 amperes.

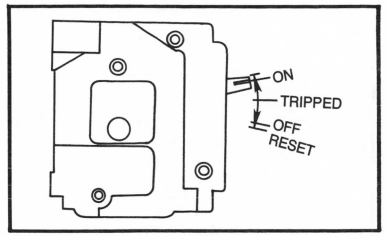

Fig. 15-4. Drawing of a typical circuit breaker showing the various positions of the handle.

This calls for a AWG #10 wire that should be fused at 30 amperes.

Sometimes the overcurrent protection size will be smaller than the rated wire size. One such example would be when the wire size must be increased above its normal current rating to compensate for voltage drop. An electric brooder draws 14 amperes of current and AWG #12 wire would normally be quite sufficient to carry the current. However, this load is 175 feet away from the power source and we find that to keep the voltage drop to within two percent AWG #6 wire is required. Table III-1 in Appendix III gives the rating of AWG #6 wire to be 55 amperes. But since the load is only 14 amperes we could use a 20-ampere fuse or circuit breaker on this circuit. Why not a 15-ampere circuit breaker? Remember we stated previously that it is recommended that certain types of circuits be rated at 125 percent above their nameplate rating. Therefore, 14 amperes × 1.25 = 17.5 amperes; a 20-ampere circuit breaker is the next closest to 17.5 amperes without going under it.

You may be wondering why we recommend oversizing the circuits by 125 percent. Besides being the National Electrical Code requirement on certain circuits slightly oversizing electrical circuits helps to cut down on excessive heat, voltage drop, etc., and allows a little room for electrical variances.

For example, take the electric brooder just mentioned. The nameplate says that at 120-volts the heater will draw 14 amperes. This meanheating element has 8.57 ohms resistance (volt divided by amperes equal ohms).

Now, it is a known fact that voltages fluctuate on power company lines during certain times during the day. One such time is during the late afternoon when most people are cooking dinner on their electric ranges. The heavy load imposed on the power lines at this time causes a voltage drop throughout the power company's lines. To compensate for this voltage drop many power companies raise their voltage at the generators at this time by possibly as much as ten percent. This is fine, but suppose for a few minutes, your 120-volt normal line voltage increases to 132 volts; this occurs daily. Let's make another calculation to see what amperage the same brooder circuit would draw now.

$$\text{Amperes} = \frac{\text{volts}}{\text{ohms}} = \frac{132(\text{volts})}{8.57(\text{ohms})} = 15.4$$

The circuit load now exceeds the nameplate rating of 15 amperes. If the circuit was fused at only 15 amperes it could mean that you would have to replace the fuse each day when it blows late each afternoon.

The subject of overcurrent protection has barely been touched in this chapter, but it is hoped that you have derived enough information to realize the important of correct overcurrent protection. If you have a given circuit fused at the recommended rating and the fuse keeps blowing, don't increase the fuse size—something else is wrong. Either the circuit is overloaded, you have a partial ground fault or similar problem. Rather than replace the blown fuses with larger ones determine the problem and correct it. Methods given in Chapter 5 should help you solve most such problems.

# Chapter 16
# Lightning Protection
# for the Home & Farm

Lightning is the discharge of enormous charges of static electricity accumulated on clouds. These charges are formed by the air currents striking the face of the clouds and causing condensation of the moisture in them as illustrated in Fig. 16-1. When the wind strikes the cloud these small particles of moisture are blown upward, carrying negative charges to the top of the cloud, and leaving the bottom with positive charges.

Or very heavy rains or other forms of heavy condensation may fall through a part of the cloud, causing one side of the cloud to be charged positively and the other side negatively to enormous potentials of many millions of volts.

When clouds, under the conditions just described, come near enough to the ground or to another cloud with opposite charges they will discharge to ground or to another cloud with explosive violence that all of us has seen at one time or another.

We say that lightning strikes various objects such as tall buildings, trees, etc. because, like all moving matter, lightning takes the path of least resistance to ground. In doing so it makes use of such tall objects projecting upward from the ground as part of it discharge path.

Rain soaked trees (especially those with rough bark), or trees with the natural sap in them have a lower electrical resistance than air. Metal buildings or those constructed of other damp materials also have less resistance than air. And the taller these objects are above the ground, the more likely they will be struck by lightning.

When any of these objects are struck by lightning the intense heat vaporizes their moisture into steam, and causes other gases of combustion that produce explosive force. And along with an electrostatic stress set up between the molecules of the material itself, heat causes the destructive action of lightning that involves a very real personal hazard and is the cause of tremendous financial loss.

## LIGHTNING RODS

Since there is a strong tendency for lightning discharges on trees, structures and other objects to travel on any metal parts which extend in the general direction of the discharge, the destructive action of most lightning discharges can be quite effectively prevented by use of properly installed I. W. D. lightning rods. In fact, an analysis of lightning fires has proved that less than five percent of lightning strikes on rodded structures has caused fire.

Lightning rods are made of copper or material that is a good conductor of electricity. They should be installed on the tops or highest points of structures or objects to be protected such as chimneys, towers, etc. On flat roofs the rods should be approximately 5 feet above the roof and ridges, and from 1 to 2 feet above the highest points on other type roofs and upward projections. The rods should also be spaced about 25 feet apart. All of the rods are connected together with a heavy copper cable of at least AWG #2 (0000 is better), with one or more ground cables of the same size run from this connection cable to the ground by the most direct path. In running this ground cable it should be as straight as possible. If any turns or bends are necessary they should be made with a long radius or gradual bends. Never bend the wire in a 90-degree bend as the lightning discharge could bypass or jump off the cable at this point. See Fig. 16-2.

The grounded end of the cable should be secured to a driven ground rod by an approved cable clamp as shown in Fig. 16-3. The rods should be a minimum of ¾-inch diameter

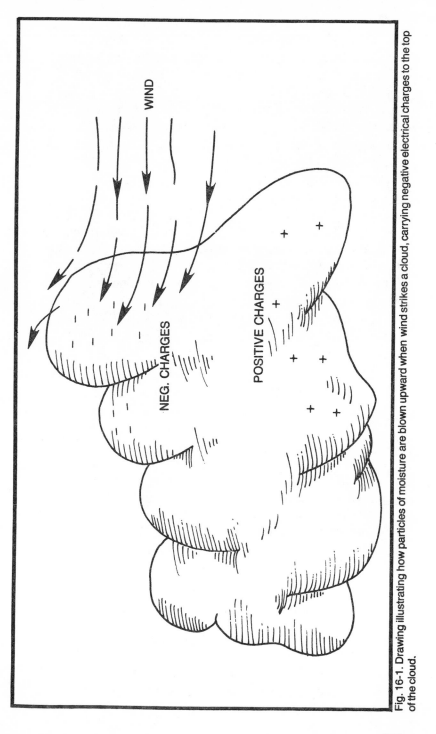

WIND

NEG. CHARGES

POSITIVE CHARGES

Fig. 16-1. Drawing illustrating how particles of moisture are blown upward when wind strikes a cloud, carrying negative electrical charges to the top of the cloud.

by 10 feet long. If the soil conditions are such that the ground resistance is high (more than 25 ohms), several ground rods may need to be driven, all connected to each other by cable, to obtain the proper ground. Always try to obtain a ground of less than 3 ohms if possible.

The tips of the lightning rods should be single-pointed for best results because it is easier for lightning to discharge to or from a pointed object than a blunt one. With the system properly installed when a direct bolt of lightning strikes the system it usually flows through the cable to ground doing very little, if any, damage to the building. This is due to the fact that the copper cable has much less resistance than wood or air and the lightning will follow the path of less resistance. In doing so less heat is also caused since the higher the resistance of an electrical conductor, the more heat it produces. A wiring diagram of a typical lightning-protection system is illustrated in Fig. 16-4.

If you attempt to install a lightning-protection system yourself (and there's really no reason why you shouldn't) the following precautions should be taken:

- Purchase lightning rods and their mounting brackets only from manufacturers whose products are approved for this purpose
- Rods should be placed on all spires, cupolas, chimneys, high dormers, gable ends, water tanks, towers, poles and similar vertical projections
- On pitched roofs install lightning rods not more than 20 feet on centers along all ridges and within 2 feet of the ends of all ridges whether they occur on the main roof or on dormers
- Interconnect all lightning rods with a copper cable not less than AWG #2 and provide at least two separate paths to ground for straight ridge-line buildings 70 feet or less in length. Additional paths should be provided for each additional 40 feet
- Install lightning rods at all corners on flat roofs and not more than 20 feet on centers around the entire perimeter. Provide additional rods spaced 20 feet on center for each 50 feet of roof width over 50 feet

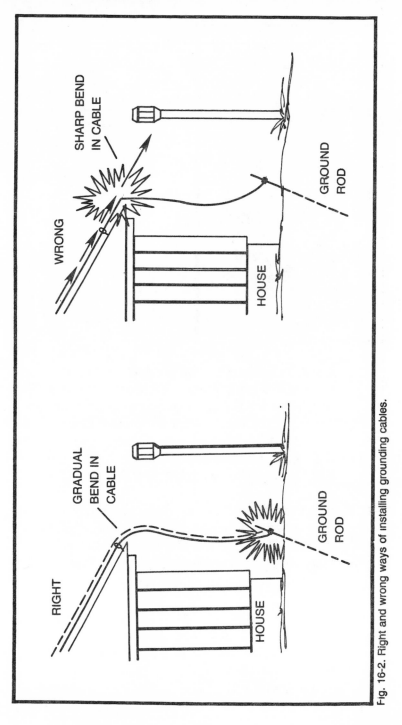

Fig. 16-2. Right and wrong ways of installing grounding cables.

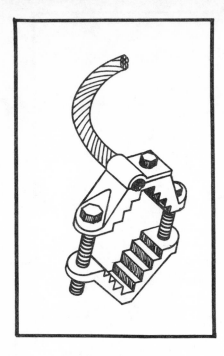

Install a minimum of two conductors to ground on roofs with a perimeter of not over 200 feet total. Install one additional conductor to ground for each additional 100 feet of roof perimeter. Figure 16-5 shows a typical lightning protection system for a flat-roofed house while you may refer to 16-4 to see a typical system on a pitched roof house or outbuilding

■ Make sure that all of the conductors run to ground are properly grounded. If they are not the entire system will be greatly affected, if not rendered useless as a lightning protective system. If there is any doubt about the gounding capabilities of your soil, have it tested with a megohmmeter.

■ All metal ventilators, guttering, electrical boxes, conduits, pipes, plumbing vents, etc. that come within 6 feet of the lightning protection conductors should be interconnected

A single lightning rod placed in the center of several objects will also protect a certain cone shaped area around the

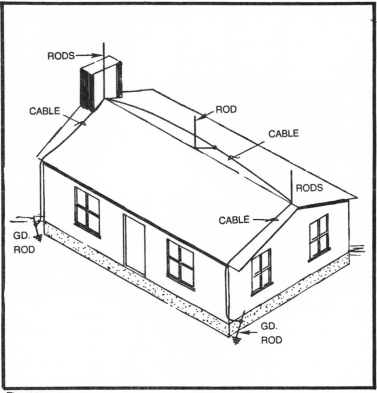

Fig. 16-4. A typical home lightning protection system.

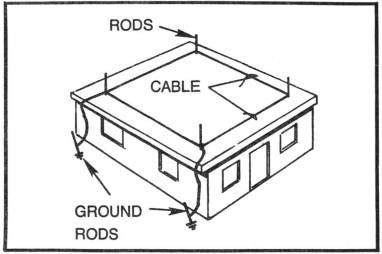

Fig. 16-5. A typical lightning protection system for a residence with a flat roof.

213

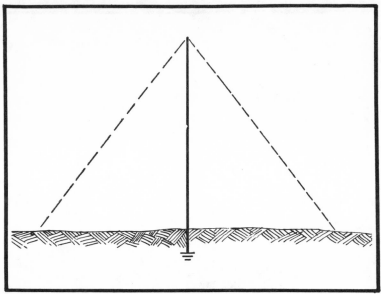

Fig. 16-6. A single rod placed in the center of an area to provide lightning protection inside the cone indicated by dotted lines.

objects as shown in Fig. 16-6. The diameter of this cone shaped area is about four times the height of the rod. Therefore a 30-foot lightning rod (above finished grade) would offer

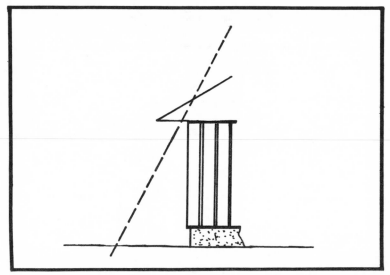

Fig. 16-7. An edge of a building roof projecting outside of the cone area as shown here will not be protected.

214

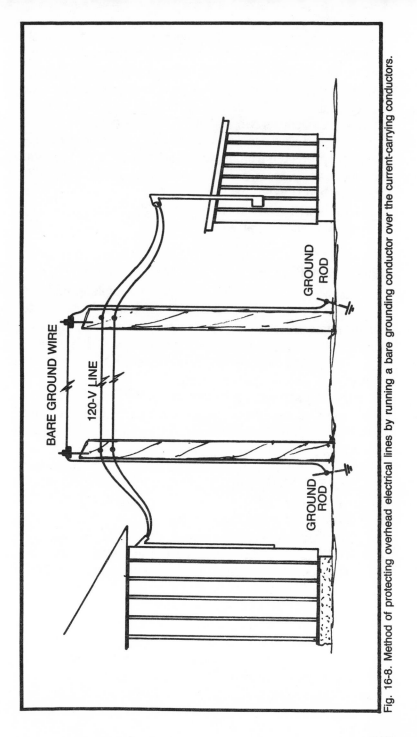

BARE GROUND WIRE

120-V LINE

GROUND ROD

GROUND ROD

Fig. 16-8. Method of protecting overhead electrical lines by running a bare grounding conductor over the current-carrying conductors.

215

protection against lightning for approximately 60 feet in all direction from the base of the rod. Bear in mind, however, that the protection is only offered inside of the imaginary cone, as indicated by the dotted lines in Fig. 16-6. So if an edge of a building roof projected outside of this cone area (see Fig. 16-7) this corner would have no protection.

Overhead electrical lines running from building to building can be protected from lightning by running a bare ground wire above the current-carrying wires as shown in Fig. 16-8. If this method is used, make sure that each end of the bare conductor is well grounded to prevent lightning charges from entering the buildings at either end.

## LIGHTNING ARRESTERS

Perhaps you don't wish to install a complete lightning protection system as just described. You would just like to provide some protection against lightning, especially on those electrical appliances that often become damaged during electrical storms, not necessarily from a direct hit but by static electricity coming in on the wires.

Lightning arresters are made in many different forms and could be the answer to most homeowner's problems. It would be well to install an expensive lightning arrester at the service drop on your home or farm electric service entrance as shown in Fig. 16-9; on pump, water-heater, and similar circuits as shown in Fig. 16-10; and TV or radio antenna as shown in Fig. 16-11. All of these circuits usually are the first to be damaged during an electrical storm.

The lightning arrester shown in Fig. 16-12 sells for around $15 and is designed to prevent lightning surges that enter through the electrical wiring from damaging interior wiring and appliances. This type of home lightning protector is a sturdy, weatherproof, service-proven device that immediately drains lightning surges harmlessly to ground. It may be installed at the service head or main switch box, or where conductors attach to other outbuildings, pumps, etc.

This type of protector discharges a lightning surge in a fraction of a second and will perform this protective function over and over again, without any maintenance or resetting the

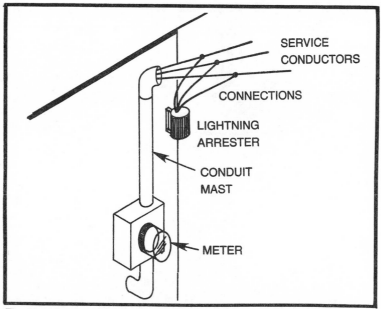

Fig. 16-9. Connecting a lightning arrester to the service entrance conductors.

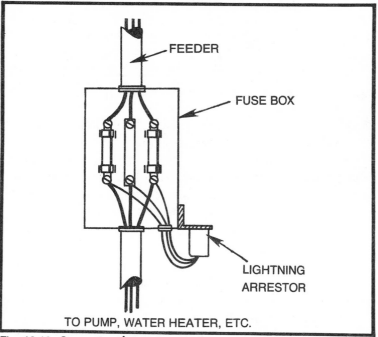

Fig. 16-10. Connection of lightning arrester to fuse box of circuits feeding pumps, water heaters and the like.

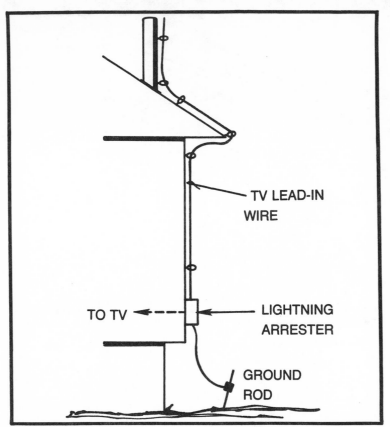

Fig. 16-11. TV antenna lightning arrester connected in system.

device. It is a two-pole, three-wire device designed for use on single-phase 120/240-volt grounded neutral service.

Each protector has three leads, two black and one white. The black wires are connected to the two hot electrical conductors, that is, the two wires that have 240 volts between them. The white wire is connected to the grounded neutral. Complete installation instructions are included in the carton as well as mounting hardware.

On two-wire 240-volt circuits feeding milking machines, coolers, pumps, and other similar electrical equipment connect one of the protector's black wires to one of the hot wires feeding the appliance and the other black wire on the protector to the other hot wire feeding the appliance. Then connect the wire to a cold-water pipe or other means of ground. The connection is usually done at the last switch box feeding the appliance in question.

Other detailed requirements for lightning protection may be found in data supplied by the National Fire Protection Association.

Lighting arrestors for protecting radio and TV antennas may be purchased from electronic supply houses, such as Radio Shack. These devices sell for less than $5 and work wonders in protecting your electronic appliances and antenna equipment from lightning and static discharge.

One type is shown in Fig. 16-13. This is designed for flat, tubular and oval twin lead or open transmission lines. The device is equipped with a sawtooth washer to grip the antenna wire securely without needing to strip or cut the wire. A ground wire is run from the terminal provided on the arrester

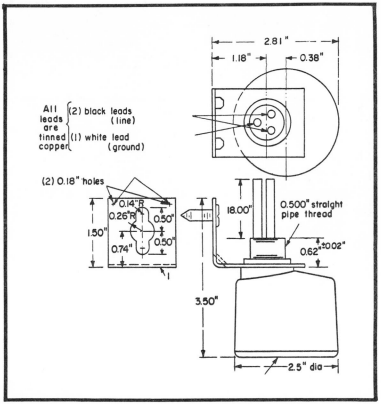

Fig. 16-12. A home lightning arrester that sells for under $15.

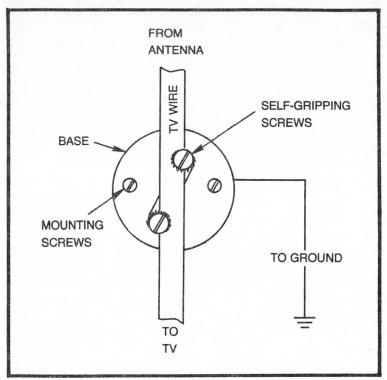

FROM
ANTENNA

SELF-GRIPPING
SCREWS

TV WIRE

BASE

MOUNTING
SCREWS

TO GROUND

TO
TV

Fig. 16-13. TV lightning arrester for flat TV cable.

and connected to a ground source such as a cold-water pipe or driven ground rod.

Other lightning arresters are available for four- and five-wire antenna rotor leads as well as for coaxial cables. All sell for less than $5.

# Chapter 17
# Television Systems In The Home

How many times have you wanted to move your TV set to another part of your home or patio and found that your present antenna lead wouldn't reach. Or if it did reach, it was probably an unsightly mess to say the least. Perhaps you live in a fringe area, miles from the closest TV station, and find that your TV reception is terrible.

This chapter is designed to show you how to correct all of these TV-reception problems and even more. You will also read how to install TV outlets in all areas of the home where you think that your TV may possibly be viewed. This includes the living room, den, family room, bedrooms, patio, and any other area that comes to mind.

For those of you who have cable TV service to your home, locating and installing the TV outlets should be your only interest in this chapter. However, there are thousands of rural residents living on farms or in small towns, who still must depend on antennas for their reception. Those are the ones that will gain the most from the instructions on improving TV reception in fringe areas.

Before actually getting into the installation of TV outlets and power-boosting components let's take a look at a typical TV antenna system in order to fully acquaint you with its operation and purpose.

Figure 17-1 shows a sophisticated TV antenna system designed for a far-fringe area up to 200 miles away from the closest TV station. Most systems, however, have fewer parts and components.

## ANTENNA HEAD

There are several antenna heads for VHF, UHF, and FM reception to choose from to take care of your reception situation, ranging from local through far-fringe areas. Most range in size from nine elements, good for a range of about 30 miles, to sixty-seven or higher elements for distances to over 225 miles. The terrain, of course, has much to do with the reception. For example, if the terrain is flat the reception range is greater than if the terrain is hilly or mountainous.

Most modern antenna heads are preassembled with fold-out elements to ease installation, and all come with complete instructions for assembling and connections.

## PREAMPLIFIER

Preamplifiers are designed primarily for use in fringe areas to boost the signal. Most come in two units. One part mounts on the antenna mast, and the other is placed near the TV set.

In general, the mast-mounted preamplifier overcomes downlead losses and rejects interference while matching your roof antenna to your TV or FM set under all atmospheric and weather conditions. An indoor-mounted AC power supply which plugs into a conventional 120-volt wall outlet sends current to the mast-mounted preamplifier. The combination of the two units amplifies all signals for channels 2 to 83 plus FM reception. If more than one TV set is to be used on the system, power-boosters are manufactured for use with four or more sets.

## ROTOR

An antenna rotor system consists of a motor-operated rotor mounted on the mast, a control unit inside the home, and a four-conductor power control cable connecting the two. The inside unit connects to a 120-volt power source for proper operation.

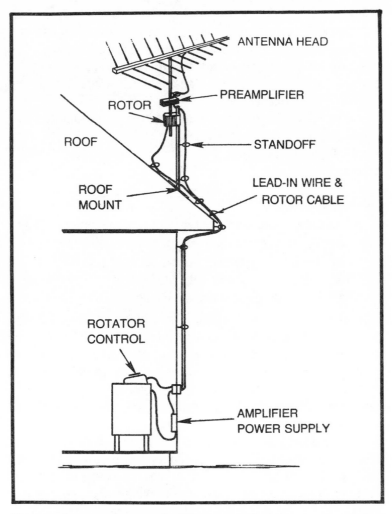

Fig. 17-1. Basic components of a TV antenna system.

Once the rotor system has been mounted and properly connected, set the inside control unit to the direction you want your antenna to point. The rotor on the mast turns the antenna to that direction and turns off automatically when the direction is reached.

Another semiautomatic type of control works very similar except that a control bar is depressed for the rotor to turn the antenna, but must be held down until the best picture is seen, then the bar is released.

## MASTS

Mast kits for TV antennas come in a variety of types from chimney mounts, through roof mounts, ground mounts secured by brackets connected to the house, to self-guying towers. Figure 17-2 shows a well-constructed tower which is mounted in a concrete base.

The height of the antenna mast is very important in order to obtain the best possible reception. If your local TV shop can't tell you the best height for your area have a test made to determine this.

## LEAD-IN CABLE

Basically, there are four types of lead-in cable; namely,

■ Twin-lead 300-ohm ribbon-type cable
■ Twin-lead 300-ohm foam-insulated cable
■ Twin-lead 300-ohm shielded cable
■ Shielded 75-ohm coaxial cable

The ribbon-type 300-ohm lead-in cable usually is considered to produce the strongest signal under adverse environmental conditions in low-interference suburban areas. However, problems occur when the cable passes near metallic objects and where high interference conditions exist as in metropolitan areas. This type is unsuitable for UHF and is considered outdated by many antenna installers.

Twin-lead foam-insulated cable is also a flat-type cable, but is totally encased in and surrounded by polyethylene foam, backed up with an outer polyethylene jacket. This type of cable offers a high resistance to ultra-violet rays, oil, fumes, moisture, salt air and abrasion. It has excellent UHF characteristics and is the type most often installed by professionals.

Another type of 300-ohm cable is the shielded type. Designed for eighty two channel color TV reception this type of cable combines the strong signal strength of twin-lead cable with the clean signal of shielded coaxial cable. The shield helps eliminate ignition and other interference caused by line pickup. A shortcoming of this type is that it is bulky and hard to handle.

The 75-ohm cable, while more costly than most other types, provides greater efficiency by minimizing interference

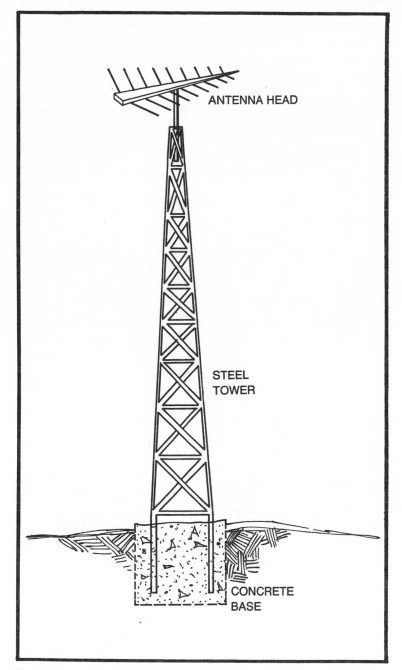

ANTENNA HEAD

STEEL
TOWER

CONCRETE
BASE

Fig. 17-2. Metal tower designed for mounting in concrete base with antenna head on top.

and is highly weather-resistant as well as being easier to install. This type of cable can be installed anywhere, even over metal objects. However, since antenna output and TV set input are 300-ohms, a step-down (300- to 75-ohm) and step-up (75- to 300-ohm) transformer must be used for the signal transfer, hence there is a large signal loss due to the many devices required. If more than one set is to be viewed a system amplifier is recommended.

## COUPLERS

When two or more TV sets are to be fed from a single antenna system, an all-channel multiset coupler must be used to divide the signals evenly between the number of sets and to eliminate interset interference. The most common types for use around the home are designed for two or four sets.

## ANTENNA SIGNAL SPLITTER

A signal splitter separates all-channel signal into individual VHF/UHF/FM signals for TV set input. The compact device mounts on the back of the TV set. The all-channel lead-in wire should be connected to the correct terminals on the unit. This is clearly marked on the bottom of the unit. Then a separate lead-in wire for each individual signal (VHF/UHF/FM) connects to three other sets of terminals on the back. Two types are common: VHF/UHF splitters and VHF/UHF/FM splitters.

This describes the basic system. Now let's start back at the top to see how to analyze each part of the system so that you can start having better reception on your own system. Again, refer to the illustration in Fig. 17-1 to determine what parts your TV system has, or what you may need to add or replace in order to achieve better TV reception.

Beginning at your existing antenna head check for bent or missing elements (signal-collecting rods). If there are ones missing or bent, your antenna cannot deliver the maximum TV signal it's designed to provide. Also check for corrosion as this can seriously affect TV reception. A corroded antenna causes your TV to receive much less signal than from a clean antenna.

Does your present antenna head have enough elements for your location? Is your antenna head mounted high enough? As mentioned previously, check with your local TV repairman for correct antenna heights in your area or have your exact location checked. This test is usually done with a field-strength meter. It can also be done with a portable battery-operated TV. As far as the correct number of elements are concerned, they will vary from manufacturer to manufacturer, but the following will give you some idea of how many elements you should have on your antenna for the distance (flat terrain) you live from the closest VHF TV station.

| ELEMENTS | MILES |
|---|---|
| 8 – 12 | 30 – 40 |
| 18 | 50 |
| 29 | 70 |
| 44 | 90 |
| 49 | 110 |
| 59 | 125 |
| 67 | 150 – 225 |

If your antenna head does not have the correct number of elements or it has elements bent or missing purchase a new one of the correct size. If your current antenna has enough elements but they appear corroded, clean them with emery cloth—gently. Remember that hilly terrain will shorten the range of your antenna. So if you live in a mountainous region, say 70 miles as the crow flies from the TV station, you'll probably need the most elements you can find for the best reception, such as the 67-element model.

Bear in mind that by purchasing a larger-than-necessary antenna may enable you to bring in more TV stations. For example, one homeowner lived within 40 miles of one TV station and his 18-element antenna gave him very good reception. When the antenna was rotated towards Washington, D.C., he was able to even pick up one channel from there, and this city was 70 miles away. However, when a new 67-element antenna was purchased, a booster added, and a rotor added to his system, he was able to pick up channels 2 through 13, which included Washington, D.C. Baltimore, Maryland, Harrisonburg, Virginia, and Richmond, Virginia, not to mention a few UHF channels also. Obviously, for a couple of hundred dollars, he was able to add much family pleasure to his home in the way of viewing TV.

Once you have assured yourself that your antenna head is the correct size and in good condition, check the antenna lead-in wire. If it is brittle or damaged replace it. If 300-ohm ribbon-type cable passes close to any metal objects use standoffs to keep the cable a proper distance or else replace the antenna cable with shield coaxial cable.

In TV reception areas where broadcasts originate from different locations an antenna rotor may help you to receive a more concentrated signal. This same rotor will also provide for optimum reception in fringe areas when supplemented with an antenna amplifier to increase the signal strength.

Just remember, your TV picture, regardless of the quality of set you may have, can be only as good as the antenna system that delivers signals to it.

## TV OUTLETS

How many times have you wished that you could move your portable TV set from your family room to the outside patio to watch the afternoon ball game? Or maybe you'd care to move it up to your bedroom for the late movie. Perhaps the next afternoon, your wife would like to watch her favorite soap opera while she's fixing the evening meal in the kitchen. Of course, you can install a long cable to drag from room to room, or you can do the practical thing and merely install TV outlets in these locations. Then all you have to do is unplug your antenna from the outlet, move the set to any other location in the home where there's another TV outlet, plug in the antenna leads along with the 120-volt power cord and you're in business.

The first step would be to determine exactly where you want the various outlets located. The second step is the easiest way to run the cable from your antenna, looping from outlet to outlet. This sounds simple, but in existing structures, routing the cable can be a difficult job if you don't want to a lot of cutting and patching of your home.

For single-floor houses with an accessible crawl space or basement under the first floor all of the cable can be fed from underneath without too much cutting, drilling and patching.

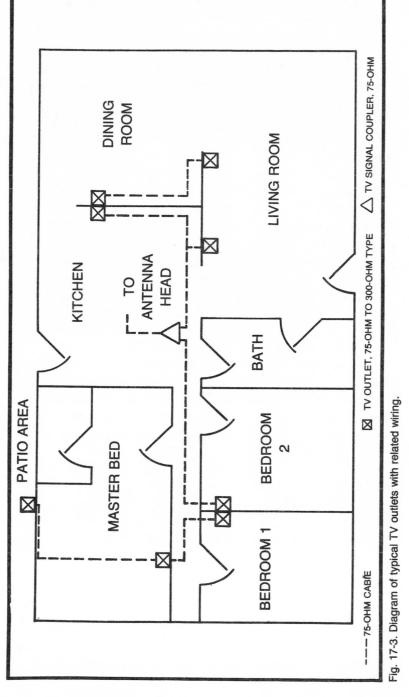

PATIO AREA

MASTER BED

BEDROOM 1

KITCHEN

TO ANTENNA HEAD

BATH

BEDROOM 2

DINING ROOM

LIVING ROOM

– – – – 75-OHM CABĺE

☒ TV OUTLET, 75-OHM TO 300-OHM TYPE

△ TV SIGNAL COUPLER. 75-OHM

Fig. 17-3. Diagram of typical TV outlets with related wiring.

229

For example, locate the outlets in the various room, cut an opening for a conventional plastic outlet box (as described earlier in this book for installing receptacles and switches), locate the point directly under the outlet box opening in the wood partition. Then drill a ¾-inch hole through the wood floor making certain that the drilling stays well within the boundaries of the wall partition in which the outlet box opening was cut. Repeat this procedure at each of the outlet locations before pulling cable from your antenna head to each of the outlet locations.

When 75-ohm coaxial cable is looped from outlet to outlet, pulled inside of each of the outlet boxes with about 12 inches hanging out of the box, install conventional 75-ohm to 300-ohm TV outlet receptacles and covers. A typical diagram is shown in Fig. 17-3.

Of course, you will need a 120-volt outlet at each of your TV outlets so try to locate the TV outlets near an existing receptacle. If this isn't possible, you'll have to install duplex receptacles where necessary as described in Chapter 6. Any TV outlet or 120-volt power outlet installed outside must be enclosed in a weatherproof outlet box.

For new construction where the partitions allow access for wiring, an outlet box mounted on a wooded stud at the desired locations may be used. Then a piece of conduit run from the outlet box to a readily accessible location after the walls are finished. A basement, crawl space, or attic are fine. See Fig. 17-4 for details.

Other accessories for your TV antenna system may include a lightning arrestor (see Chapter 16), a filter to eliminate any antenna-fed interference causing picture distortion, etc. Complete details on these accessories may be had at your local electronic shop or found in some of the large electronics mail-order catalogs.

## MASTER ANTENNA TV SYSTEM

A small master antenna TV (MATV) system is shown in Fig. 17-5. It can supply eight outlets with an equal amount of TV signal even if some of the outlets are not in use. Antenna systems and TV signals differ from electricity in that even if the outlet is not in use the service drop itself draws signal from

230

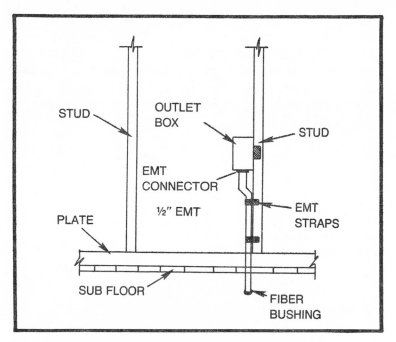

STUD

OUTLET
BOX

STUD

EMT
CONNECTOR

½″ EMT

PLATE

EMT
STRAPS

SUB FLOOR

FIBER
BUSHING

Fig. 17-4. TV outlet box detail for installation in new construction.

the system. This is designed into the system so that when TVs are switched on and off they don't affect the other receivers on the same antenna.

Normally TV couplers are available in either two-set and four-set varieties. Each time one of these devices are inserted in the antenna system the signal is reduced, the four-set type causing the greatest drop in signal.

The first step to do when installing a system is to put up the antenna head. A survey around the neighborhood will give you a clue as to what size of antenna head to purchase. Don't be afraid to ask around first because you can learn by other peoples mistakes and good fortune. If you have a modern color receiver with UHF and there are some UHF stations nearby then purchase a VHF/UHF antenna. But if there are no such stations within 11 miles forget it and save yourself the dollars.

Don't buy an amplifier until you are sure that it will help. That is, if you have installed a system without the optional amplifier and the signal is snowy you may or may not need an amplifier. To determine if an amplifier will help connect a receiver to the antenna lead-in before it reaches the two-set coupler. If the reception is good there then an amplifier will help. But if it is snowy there too then an amplifier will only amplify the snow even more. In this case you might try an antenna-mounted preamplifier, a larger antenna head or a higher mast or tower.

# Chapter 18
# Music All Through the House

Before you begin throwing hundreds of dollars into a built-in stereo music/intercom system for the home, whether for new construction or not, it is important that you carefully and thoroughly plan the system. For new construction the location of the speakers must be chosen as well as their mounting height. You must also make certain that a cabinet or piece of furniture will not later be in the way of any speaker or controls. When installing a music system in existing homes, you will also have to consider the routing of the wires in order to cut down on unnecessary cutting and patching.

## CAREFUL PLANNING

The first consideration should be the various components available for the system; then determine the relative importance of them to you. You will have a choice of at least ten different components, including AM and FM radio, TV audio, record changer, three tape decks, intercom (including talk-listen door speakers), door chimes and security/fire alarm system.

Since most of the systems will be built into the structure, we cannot emphasize too strongly the importance of very careful planning. We have said this before, and we will say it several more times during the installation details given in this chapter.

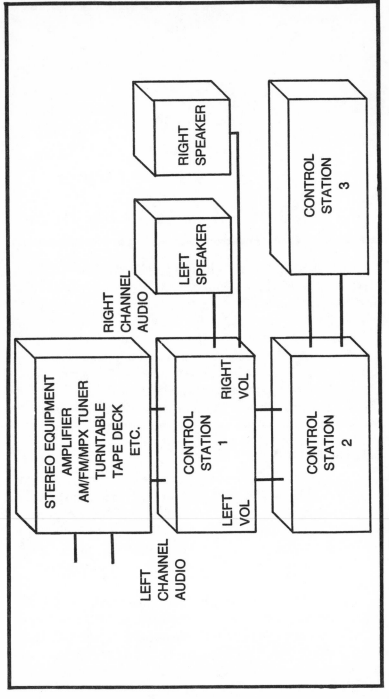

Fig. 18-1. Pictorial diagram of a stereo music system.

234

A pictorial wiring diagram of a stereo music system is shown in Fig. 18-1. The master unit in this diagram is designed to be built into a wall and contains AM/FM radio, FM stereo, record changer, and a tape deck to record anything coming through the system. Or use extra microphones to record parties and treasured family get-togethers. Most of these units are sized to fit between standard partition studs (14½ inches) and are therefore easily installed in either existing or new partitions provided rough lumber was not used for the studs. In this case, you may have to chisel a half-inch or so from either side of the studs in order to make the unit fit. Figure 18-2 shows the dimensions of a typical rough-in frame for a master unit.

## INSTALLATION

If the master unit is mounted in a vertical position as shown in Fig. 18-2 mount the rough-in frame approximately 25 inches from the finished floor to the bottom of the frame. For a horizontal arrangement (Fig. 18-3) mount the unit about 4½ feet from the floor to the bottom of the unit.

During this rough-in stage, you will also have to provide a 120-volt AC power supply, an antenna lead-in cable for the

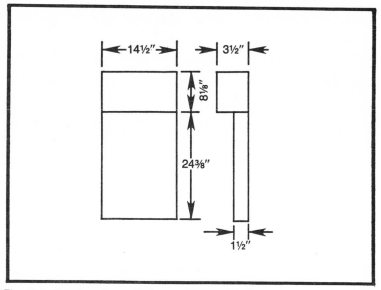

Fig. 18-2. Drawing of a vertical-mounting rough-in kit.

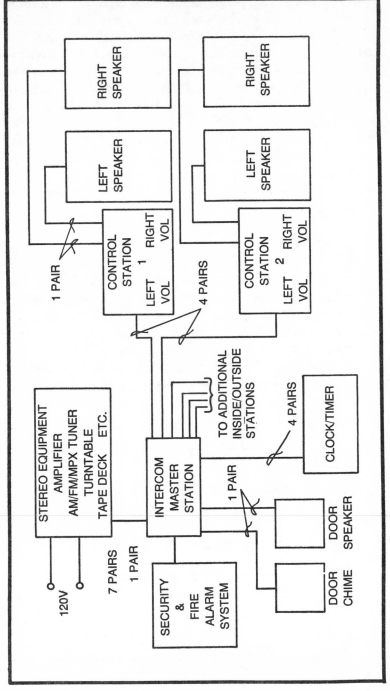

Fig. 18-3. Drawing of a stereo system with components mounted in a horizontal position.

FM system (see Chapter 17) and speaker and control wiring. In the music-only system, two pairs of AWG #18 wires usually will suffice, but for a stereo/intercom system like the one shown in Fig. 18-4, you will need seven pairs of AWG #18 wires running to the intercom master station.

Speakers may be either ceiling or wall mounted, but all speakers used outdoors—on walls or soffit—must be weather resistant. Use two for stereo and one for monaural. Stereo speakers should be installed within the same wall approximately two-thirds the wall length apart, but never closer than 4½ feet apart or farther than 15 feet apart. Each set of speakers should be controlled by a remote volume control located in or near the area where the speakers are located.

Although the exact wiring of the system will vary with each manufacturer, the following will illustrate a typical application (Fig. 18-4).

Interconnect the master unit to the intercom master unit with 7-twisted-pair cable before running individual 4-twisted-pair cables from the intercom master to remote

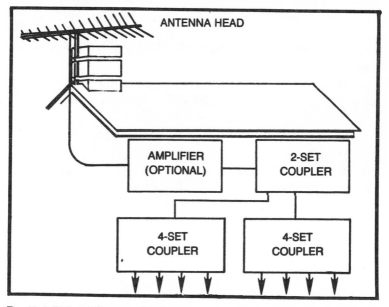

Fig. 18-4. Pictorial diagram of a stereo system with intercom, electronic chimes, security system, and others.

control locations. Make the connections as indicated in the instructions provided with each system.

Besides providing the luxury of AM/FM radio anywhere in the home this system includes a family message center, a cassette tape player/recorder, a digital clock, an 8-track tape player, and a record changer...to name a few.

The easy-to-operate cassette tape player/recorder can record from the system's radio, the auxiliary record changer or the 8-track tape player. As the family's message center it offers special controls for recording and playing back messages to and from family members through the master speaker. A signal in the form of a light tells at a glance when a message is waiting for a playback.

The digital clock provides the time at a glance in illuminated, easy-to-read numerals while a security/fire alarm system connected to the master unit provides extra peace-of-mind as they can detect lethal smoke, heat from fire and even attempted forced entry.

Electronic door chimes relay chime tones through the intercom system and if you can't answer the door right that minute, you can acknowledge your guest by means of the built-in door speaker. Since the door speaker is controlled by remote control it allows your callers to answer hands free. Various speakers are available for mounting in either wall, ceiling, or soffit.

Begin the installation of this type of system by selecting the preferred locations for the master station, each remote station, door speaker, remote control units, and the alarm devices as described in Chapter 20. For maximum operating convenience, install the master and inside stations 4½ feet from the finished floor to the bottom of the unit. Make certain that the master unit is located at least 4 inches from adjacent walls, cabinet or countertops.

Whenever possible try to locate the inside units on interior walls that are free of insulation and other obstructions. In order to prevent feedback or interference, never mount speakers or any other of the controls or devices back to back or in a common wall between rooms. Door speakers, of course, will have to be located on an exterior wall adjacent to the entrance door or else located in a nearby porch ceiling or

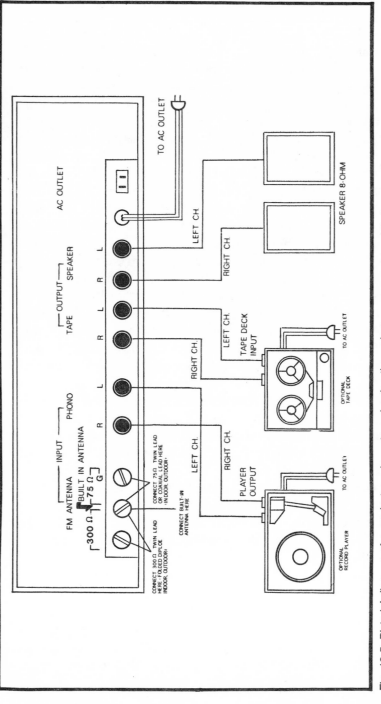

Fig. 18-5. Pictorial diagram of a modest music system showing the various components and their connections.

239

soffit overhang. Again, all exterior components must be of a weather-resistant construction.

An FM antenna should be provided for the radio if you cannot use your existing TV antenna. Chapter 17 gives all the details about connecting an FM set to a VHF/FM antenna head.

Notice in the wiring diagram in Fig. 18-4 that individual flat parallel ribbon-type four-wire cable is run from the master unit to each inside speaker or remote control location. The power transformer should be located near the panelboard supplying the home with 120/240-volt electricity. However, if it is not feasible to locate the transformer near the electric panel make certain that the location you select is readily accessible. Then connect the primary side of the transformer to a 120-volt circuit. Two-conductor, AWG #18 wire connects to the load side of the transformer (secondary side) and extends from the transformer to the master unit. The entire system should be properly grounded to an earth ground.

If you don't want to go into your stereo music system quite as elaborate as previously described, there are, of course, simpler outfits available. For as little as $250 you can purchase and install a built-in stereo system and still have a versatile unit like the diagram of the one in Fig. 18-5.

A system of this sort contains an AM/FM/MPX radio with an 8-track stereo tape player. This is the basic receiver unit and requires conventional 120-volt house current as its power source. However, before connecting the power source and turning the set on, make certain that the speakers are properly connected, otherwise damage may occur to the amplifier.

Most of these self-contained stereo music systems have a built-in antenna already connected to the set at the factory. However, if the built-in antenna does not give satisfactory results, try an alternate antenna arrangement like one of the following:

1. Fully extend the 120-volt line cord to insure that the built-in antenna is providing its peak performance.
2. Disconnect the built-in antenna and connect a T-shaped (dipole) antennas to the 300-ohm terminals on the back of the set.

3. Disconnect the built-in antenna and connect an indoor TV/FM antenna to the 300-ohm terminals.
4. Disconnect the built-in antenna and connect an outdoor TV/FM antenna. If the leads are 300 ohms or 75-ohm coaxial cable with a 75- to 300-ohm transformer on the end, connect to the 300-ohm terminals, or if you prefer to use only the 75-ohm cable with split connectors, connect to the 75-ohm terminals as shown in Fig. 18-6.

Referring again to the diagram in Fig. 18-5 notice that other components and devices may be connected to the basic unit. A record changer and a tape deck (both requires a 120-volt power source) are two. Then you can plug in an electric guitar or a microphone through a front-panel jack. Or if you don't want to disturb others while playing or listening, a set of stereo headphones may be plugged into the set so only you can hear what's going on.

When you are investigating the various models on the market the following terms may help you to better understand what you are buying.

master station—The basic unit of every intercom system. Normally houses the radio, amplifier, intercom circuitry and its own speaker.

remote station—A speaker located away from, and connected to, the master station.

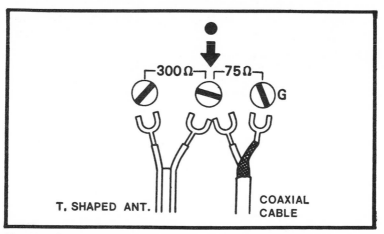

Fig. 18-6. Detail of antenna connection on a typical stereo tuner.

**door speaker**—A remote station used at outside entrances for talking to the master station or remote stations. Most are designed specifically for intercom use rather than for radio or music reception.

**centralized system**—The master station includes switches to control the function or operation of the remote speakers; that is, to turn the stations on or off and to monitor.

**decentralized system**—The master station includes switches selected by controls on each remote station or remote control panel. Therefore, no switches are provided on the master unit.

**FET**—Field-effect transistor. Its main advantage over the standard type is higher gain—better reception results when used in AM/FM/MPX receivers.

**master all-call/all-listen**—Lets you page through all speakers at once from the master. Music is automatically silenced. Person called can answer without operating speaker controls.

**station all-call/all-listen**—Intercom calls from any remote speaker as well as from the master can be answered hands free without having to operate any controls.

**door answering**—An outside visitor operates a talk/listen switch on the door speaker to talk and reply to remote stations.

**hands-free door answering**—Permits visitors to talk from the door speaker without having to operate any controls. In systems having this feature door speakers do not have talk/listen switches.

**electronic door chime**—Includes electronic circuitry to relay chime sounds through all speakers. Can be heard above music or intercom.

**intercom override**—Intercom messages will be heard above music.

**monitor**—Stations can be set to listen-in and relay sounds from its locations to the master or other stations. Sounds will be heard over music.

**music muting**—Music is automatically silenced during intercom calls.

**privacy**—Volume control adjusts to turn remote station completely off.

**room-to-room intercom**—Lets you talk to and receive calls from remote and master stations. Remote stations however cannot be selected individually from other remote stations.

**standby intercom**—Provides intercom without music at remote stations.

# Chapter 19
# Notice Your Guest

One of the simplest and most common electric signal system is the residential door-chime system. Such a system consists of a low-voltage source, a pushbutton, bell wire and a set of chimes. The quality of the chimes will range from a one-note device to those which play lengthy melodies.

The modern door chime has a plunger which strikes against a bar or hollow tube to produce a pleasing musical note. Figure 19-1 shows a typical wiring diagram of a two-note chime controlled at two locations. One button at the main entrance will sound the two notes when pushed, while the other button at the rear door will sound only one note when pushed. The components of this circuit consists of:

■ **pushbuttons**—The typical residential pushbutton will have two contacts (Fig. 19-2) and acts very much like the common single-pole wall switch used to controlled lighting.

■ **transformer**—Chime transformers are made for connection to 120-volt lines in order to reduce the voltage to 10 to 16 volts for the operation of low-voltage chimes. A 120/10-volt 5-watt transformer will be quite sufficient for all single-note chimes. A 120/16-volt 10-watt transformer is recommended for two-note chimes, while a 120/16-volt 15-watt

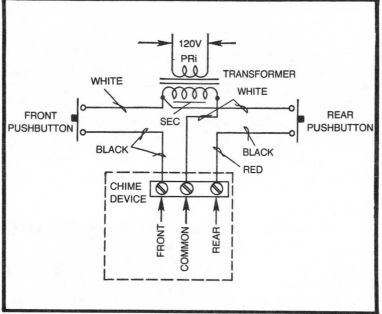

Fig. 19-1. Typical wiring diagram of a two-note chime controlled from two locations.

transformer is recommended for multiple chime installations. If the correct chime transformer is selected for the application, it should last a life time. Figure 19-3 shows a typical chime transformer.

■ **wire**—Bell wire cable, consisting of two or three wires, is normally used for all low-voltage wiring including wiring for residential chimes. Its small size makes it easy to install either concealed in the walls or run exposed like on the top of a baseboard.

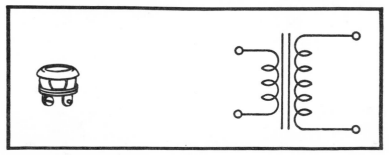

Fig. 19-2. One type of residential pushbutton.

## LOCATION OF CHIMES

In new construction, the chimes should be located where they can be heard in all areas of the home. In existing construction, however, it may be impractical to run the wires to a location which is ideal. In this case try to locate the chimes where they can be heard in *most* areas of the home.

Larger homes may require two or three chimes connected in parallel. They will respond to the same pushbutton at the same time. Or you can use an electronic chime that sounds through your radio-intercom system as described in Chapter 18.

## TYPES OF CHIMES

Unlike the unattractive door bell of yesteryear, modern chimes are available to fit nearly every purpose and decor. For example, besides the wide variety of different styles to fit any decor clock chimes are also available. Now there is no need to conceal the chime in a dark hallway or behind a door. It may be used as a decorative focal point or as part of a wall treatment in the kitchen, family room, living room or upstairs hallway. Even nonelectric models are available for fastening directly to your door in case the running of wires (such as in an existing structure) is impractical.

There are also an array of pushbuttons, lighted and unlighted, to grace the doorway of any home regardless of its architectural style.

## INSTALLATION OF A TWO-NOTE CHIME

Once you have decided upon the style and type of chimes you want in your home and have determined the best location the selection of the type and quantity of incidental materials is the next step. Besides the chimes, you will need a low-voltage transformer to reduce the 120-volt house current to 10 or 16 volts. In the case of a two-note chime a 16-volt 10-watt model is recommended. Two pushbuttons, rated at 16 volts of a style agreeable with your decor and taste will be needed as well as enough two-wire AWG #18 bell cable to reach from each pushbutton location to the chime itself. A piece of three-wire AWG #18 bell cable long enough to reach from the

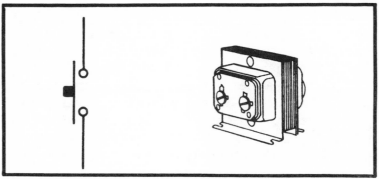

Fig. 19-3. A typical chime transformer.

transformer to the chime and some wire staples should complete the material list.

### Mounting the Transformer

Locate an accessible lighting outlet box near to where the chimes are to be mounted. In new construction or where the building structure is open any outlet box near the chime location will suffice. However, in existing homes where the structure is concealed the outlet box must be easy to get to as well as be in a location conducive to running the bell wires with the least amount of cutting and patching.

If the house is a one-level home usually the transformer can be mounted on a nearby outlet box by getting to the box from the attic. If a two-story house is the case and the chimes are to be mounted on the first floor an outlet in the basement or crawl space would be the best choice.

Pull the fuse or trip the circuit breaker providing overcurrent protection to the circuit feeding the outlet box. Next, remove the lighting fixture, wiring device, or blank plate attached to the outlet. Then locate a vacant knockout plug on the side or top of the outlet box and remove this as described previously. You are now ready to connect the transformer.

There are two basic types of low-voltage transformers: one is made to mount directly onto the outlet box (see Fig. 19-4) and the other is mounted remote from the box (Fig. 19-3). In most cases the type mounted inside the box makes a better and safer job, so try to purchase this type if at all possible.

247

If the direct-mounting type is used insert the two lead built into the transformer through the half-inch knockout opening on the outlet box. Also place the round attaching speed nut through the knockout so that the securing screw on the transformer may be tightened to secure the transformer in place. Some care is required here because the nut usually has a shoulder that must be aligned before the nut can be tightened.

When using the remote-type transformer secure the base at the nearest wood structure to the outlet box. This could either be a joist, stud, bridging, etc. Then use an NM connector in the half-inch knockout, feed the two leads from the transformer through the connector and tighten, but not tight enough to damage the insulation on the wires.

## Connecting the Transformer

Leave the current off in the dead outlet box in which the transformer leads are located. Secure one of the transformer leads to a white neutral wire with a wire nut. Make certain you have neutral wire and not a switch leg controlling the lighting fixture attached to the box. Next secure the other transformer lead to a hot (black or red) wire. Again, make sure that this is not a switch leg because if it is, the chime will only work when the light fixture is turned on. Secure the splice with a wire nut and replace the lighting fixture or other device that was attached to the outlet box originally.

You may now activate the circuit as the only exposed parts of the transformer will be the secondary terminals which will have only 10 to 16 volts; this will cause no harm or shock.

## Pulling the Wire

In general, the two-wire cable from the pushbutton locations to the chimes should be run in as direct a route as possible, and in a manner so that the wire will be protected from nails (driven later) and other possible dangers. When securing this wire with staples make certain that you don't drive the staples in far enough to cut or otherwise damage the insulation on the bell wire cable.

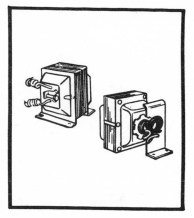

Fig. 19-4. Low-voltage transformer designed for mounting directly to outlet box.

When installing the cable in existing homes methods of fishing the wire may be employed as described earlier in this book.

As mentioned previously, you will need a two-wire cable run from each door or pushbutton location to the location of the chimes, and a three-wire cable from the chime location to the transformer. Make certain that you leave enough at each termination point to make the splices and connections.

## Installing the Pushbuttons

There are many types of pushbuttons. Some are surfaced mounted, others may be recessed or semirecessed. The exact installation procedures will differ with each type. However, instructions are usually included with each type to insure proper installation of them.

One type is installed by first drilling a half-inch hole in the outside door trim at approximately 4 feet above the finished floor. The hole should be about 1½ inches deep. Then, with a quarter-inch bit, drill another hole on through the timber to the void space within. The bell cable can be fished up (or down) in this void space, then pulled out through the quarter-hole into the half-inch hole. Strip the ends of the cable and connect the end of each wire onto the terminal on the pushbutton. With your thumb press the round pushbutton assembly (made exactly to fit the half-inch hole) into the hole until the trim is flush against the door trim. Figure 19-5 shows a cross section of this installation.

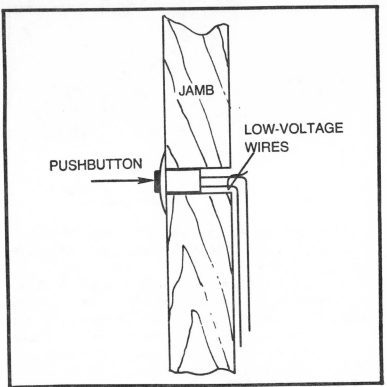

Fig. 19-5. Cross section of a flush-mounted door pushbutton.

## Making Final Connections

Figure 19-1 shows how the connections are made. Both two-wire cables from each pushbutton are run to the low-voltage transformer. The white wires in these two cables are then spliced together and connected to one of the terminals on the secondary side of the transformer.

The black wire from the front door pushbutton is connected to the black wire of the three-wire cable running from the chimes. The black wire, of the rear door pushbutton is connected to the red wire of the three-wire cable, while the remaining white wire is connected to the other terminal on the transformer.

Markings on two-note chimes may vary, but all of them will have three terminals; probably marked $F$ for the front door or two-notes, $R$ for the area door or one-note, and $C$ for the common terminal. Therefore, the black will connect to the

$F$ terminal, the red wire to the $R$ terminal and the white wire (from the transformer) to the $C$ terminal.

When all of the connections are made check for correctness. If you have the buttons reversed merely reverse the red and black leads at the chime terminals.

# Chapter 20
# Security Systems

Burglary and fire pose a direct threat to everything of value to you. While nothing can prevent them from happening a properly installed security system can provide early warning of these threats before it's too late, assuring you extra peace of mind and a more comfortable feeling about your home and around your farm.

Of the more than 12,000 lives claimed by fire each year, more than half of these deaths were caused by smoke inhalation, before flames ever reached the victims. While not as much a threat to life as fire, burglaries have caused property loss that run into the millions each year.

You cannot be too careful. In this chapter, you'll find how to install a fire alarm system capable of sniffing out smoke and sounding an alarm as the first wisps appear, giving you and your family precious time to escape from any danger. If an intruder attempts a forced entry in your home or farm buildings a security system described in this chapter, when properly installed, will set off a piercing sound to scare off the intruder and to warn you of his presence so you can call the police.

The devices available for home security systems range from a simple plug-in smoke detector to sophisticated electronic components that provide intruder detection circuits and alarm, fire alarms, AM/FM radio, room-to-room intercom, door chimes, and others in one system. A brief description of such a system was described in Chapter 18. But let's take the simpler devices first.

## SMOKE DETECTION ALARMS

This type of device is an easy way for the homeowner to provide his home with a smoke detection alarm system that could provide precious life-saving seconds in case of a fire. Several of these devices have been available to consumers over the past few years. The early type was powered by the 120-volt line coming into the home. The fire alarm sounded at the first signs of smoke, but if the electrical system was damaged, as is the case with many home electrical fires, the smoke detector was of no use since there was no power for it. Later models are powered by batteries, usually the 9-volt type used in many low-cost portable radios. Since they do not depend on household current they are much more reliable. However, it is necessary to check the battery condition from week to week if the system is to be of value when it counts.

If this is the type of unit you want to install in your home begin by determining the number of units you'll need. In general, one unit should be located in the hallway or other access to bedrooms. If your home has bedrooms upstairs one should be located at the top of the stairway. When the bedrooms are scattered throughout the home more units will be needed (see Fig. 20-1).

Once the number of units needed are purchased they may be surface mounted on either the ceiling or on a wall (as high as possible) by means of a bracket and screws furnished with each. This bracket is designed to attach directly to the ceiling or wall surfaces without any other provisions being necessary. When wall-mounted, the top of the unit should be not less than 6 inches and not more than 12 inches from the ceiling/wall junction. Ceiling-mounted units should be located in the center of the selected area.

Although the operating principles will vary from manufacturer to manufacturer, most contain a special type of electronic sensor that activates a horn alarm.

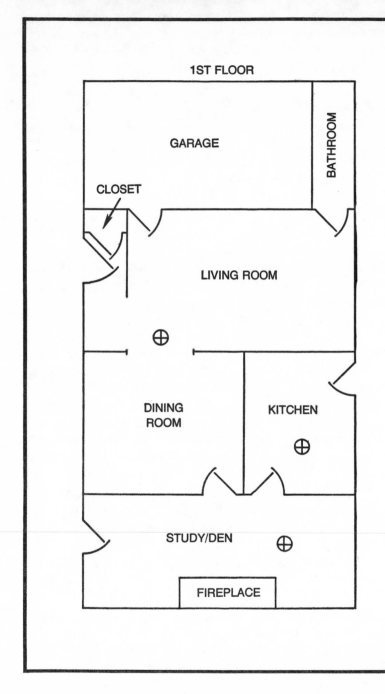

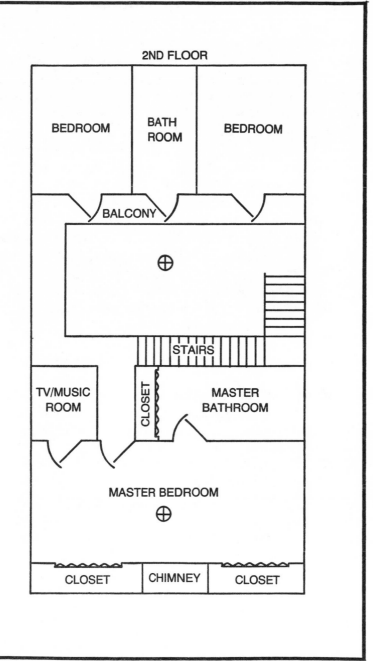

Fig. 20-1. Floor plan of a two-story dwelling. Circled crosses indicate battery-powered fire/smoke alarm locations.

## SURFACE-MOUNTED FIRE/SECURITY SYSTEMS

This type of system is well adapted to existing structures where a built-in system may not be feasible. Even the wiring can be installed in surface molding if it is not possible to conceal it in ceiling and partitions. Figure 20-2 shows a wiring diagram of a typical security fire alarm system of the type just mentioned. This system is designed to accommodate the following alarm combinations:

■ An outside alarm and flashing beacon that combines both audible and visual signals. When mounted at the high point of the roof, gable or an outside wall the entire neighborhood will be alerted should the alarm sound. Many people mount this device to their TV antenna mast for even better visibility. Usually a high-intensity inside horn (same as the outside horn without flashing beacon) is also surface-mounted in some out-of-the-way location inside of the home.

■ Two of these inside horns may be mounted or connected to the system.

■ One outside alarm and flashing beacon and two nonhigh-intensity horns may be connected together.

■ One outside alarm, one inside alarm horn, and one inside alarm horn with electronic circuitry to relay alarm signal through any radio-intercom system as described in Chapter 18.

■ One outside alarm and two inside alarm horns with electronic circuitry

Any combination of the above components may be connected to the system in Fig. 20-2 up to a total of four. The power unit must be wired into a 120-volt AC circuit or to any nearby convenience outlet with a plug-in cord set with safety lock. The power unit's features include a pilot light that tells when the system is working, circuits are monitored by solid-state circuitry at the end of the detector line, a reset button restores the system to normal operation after the alarm is tripped and a test button is provided to test the system manually. Types designed for residential use are very small, 9 inches high by 4 inches wide by 3 inches deep.

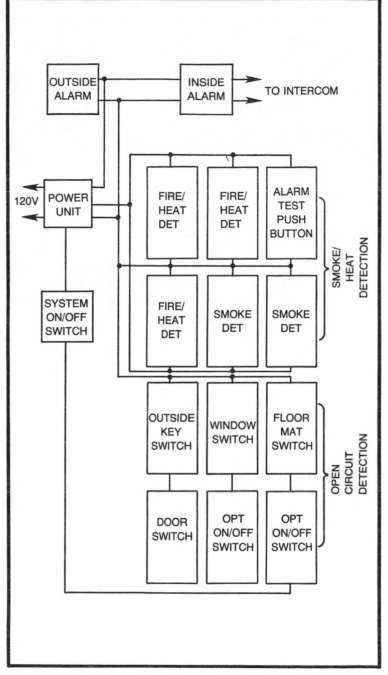

Fig. 20-2. Wiring diagram of a surface-mounted security system.

Fire/heat and smoke detectors may be connected to the system as shown in the diagram. The smoke detector automatically activates the alarm when smoke accumulates. The fire/heat detector automatically activates the alarm whenever temperatures rise above 135 degrees Fahrenheit. Any number of these devices may be connected to the system. One is recommended for every 20 by 20-foot of floor space; that is, the device will sense heat for 10 feet in all directions. Many homeowners mount them in all living areas, bedrooms, closets, etc.

A fire/heat detector mounted in areas like attics, furnace rooms and similar areas where the temperature is higher than normal, must be calibrated to sound an alarm at a temperature higher than in the living areas of the home. One model on the market is calibrated to sound the alarm when the ambient temperature rises above 200 degrees Fahrenheit.

Any number of plunger-type entry detector switches that recess in a ¾-inch hole about 1¼-inches deep may also be used. They are connected to the open circuit detection line and are usually inserted in wood window or door frames. If the recessed type is not practical a surface-mounted magnetic-type entry detector may be used instead.

But how does the homeowner get in his own home without setting off the alarm? Easily. An outside key-operated entry alarm switch is mounted near the entry door as shown in Fig. 20-3. This switch when combined with a special door cord allows you to turn the alarm circuit on or off to permit entry without sounding the alarm.

There are also inside alarm shunt switches that surface mount beside windows, entrance doors, etc. to permit the opening of windows and doors without sounding the alarm.

Where it is not practical to install fixed mounted entry detectors you may use any number of floor-mat entry detectors with the system. They are usually furnished with at least 6 feet of lead wire and may be concealed under rugs at doors, windows, stairways, etc.

You may want to install an alarm pushbutton on the wall near to your bed to manually activate the alarm. Any number of these may also be used on the system.

Fig. 20-3. Mounting locations of key-operated entry alarm shunt switches.

The exact wiring connections, methods of mounting, etc. will vary with each manufacturer of the equipment, but, in general, the following suggestions should be helpful regardless of what type system you may want to install.

The control unit may be mounted at any convenient out-of-the-way location where it will be easily accessible to responsible members of the household, but not to children or visitors.

Outside alarm devices should be mounted at some high point, such as near the top of a roof gable so that it can be seen by as many people in the neighborhood as possible. Bell alarms may be mounted on any outside wall, but they should be positioned so they are as inconspicuous as possible. The inside alarms should be mounted where they can easily be heard from the bedrooms with the doors closed.

Fire/heat detectors have an effective protection area of up to 20 feet by 20 feet when mounted on a smooth surface ceiling. If any one room as area exceeds these dimensions use more than one detector in the area. They should be located in halls, closets, bathrooms, basements, and other living areas. They should be calibrated to sound the alarm if the ambient temperature exceeds 135 degrees Fahrenheit. In areas where the ambient temperatures are normally higher than the living area (attic, furnace room, etc.) calibrate the detectors to sound the alarm if the ambient temperature exceeds 200 degrees Fahrenheit.

Smoke detectors are normally located in halls, stairways, and other approaches to bedrooms. However, install as many additional detectors as you feel is necessary to provide full warning coverage.

Most security systems are wired with AWG #18 limited energy cable or AWG #18 single conductor wire for low-voltage wiring of all components on the system. The actual installation of the wire or cable is very similar to the methods described in this book on low-voltage switching and wiring of door chimes. Always refer to the installation instructions with the type of system you purchase for exact connection details.

Figure 20-4 gives recommended points of installation for entry-detection devices for doors and windows. Note the location of the mat detectors concealed under a rug at the window and door. The remaining devices are either recessed or surface-mounted plunger-type or magnetic-type entry detectors for open circuit detection lines. Shunt switches, as described previously, are also shown in the diagram.

With a system installed in your home as discussed so far you would certainly have a better feeling about your possessions. But then suddenly, another thought hits you. What happens when there is a power failure? You have heard that this is when many burglars go into operation. Although a power outage occurs very infrequent these days, you can obtain units that have a battery standby feature if the thought of a power outage concerns you.

Nutone Division of Scovill manufactures such a system. It operates very similar to the ones previously described; that is, it contains independent fire and security circuits to provide different alarm signals through a maximum of one inside and one outside, or two inside alarm devices. A steady signal is sounded for fire while a pulsating signal is sounded for forced entry. However, if there is a power failure in the 120-volt power line, the system automatically switches to an auxiliary battery power pack for added security. A built-in battery charger keeps the battery charged when operating on conventional 120-volt house current, and a battery test feature is also included for periodic battery tests.

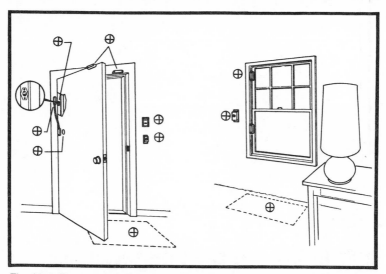

Fig. 20-4. Recommended points of installation for entry-detection devices for doors and windows, noted by circled crosses.

An additional feature that is available on some security systems designed for the home is a time-delay exit/entry feature. As the name implies, a solid-state electronic device delays the alarm, which allows you and your family to leave and enter the home without setting off the alarm. This eliminates the need for an outside key-operated switch.

# Chapter 21
# Built-In Central Cleaning Systems

While not usually thought of as an electrical system the central vacuum cleaning system is an electrically operated appliance that is installed in many residences (both existing and new) to make household chores much easier.

The power unit is installed in a central location within the home, usually in the basement, the garage or a closet and is powered by an electric motor from 1 to 2 horsepower. Disposable soil bags are used which are easily replaced when full. A complete loss of vacuum tells the user that the bag needs replacing.

Automatic inlets are located strategically located throughout the home so that the 25-foot flexible cleaning hose can reach any point within the house. Two-inch-diameter tubing is run from the central unit to each of the inlets. When the inlet plate is raised to attach the flexible hose the central unit comes on allowing the user to clean anywhere in any direction within 25 feet of the inlet.

An exhaust from the central unit to the outside is also recommended but is not necessary for efficient operation. This is just an extra precaution to guard against odors that may accumulate in the dust bag.

This chapter informs you exactly how to lay out and install a complete central vacuum system for his own. The

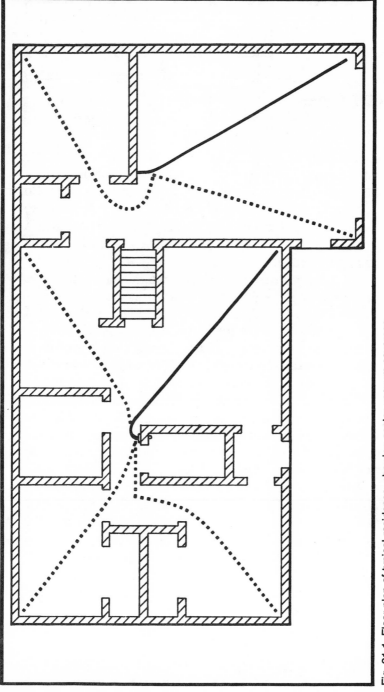

Fig. 21-1. Floor plan of typical residence, showing center vacuum coverage.

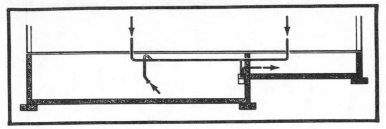

Fig. 21-2. Section of residence in Fig. 21-1.

first step is to lay out the hose inlets. Since hose lengths for most residential units are either 25 feet or 30 feet in length a radius of the same distance can be cleaned from one inlet. So with a cord the same length of the hose, locate the inlets so that every area in the house can be reached. The floor plan in Fig. 21-1 shows a typical residence where only two inlets give complete coverage.

For a uniform and neat appearance the inlets should be installed the same height as the duplex receptacles, TV outlets, etc. (14 to 15 inches above finished floor to bottom of box). They should also be located near doors or in halls so as to avoid being covered by furniture. If a stairway is present in the home try to arrange the inlet location so that the hose will be pulled up the stairs and not down for safety reasons.

A section of the floor plan in Fig. 21-1 is shown in Fig. 21-2. This section shows the power unit mounted in the basement, a utility inlet for cleaning the basement area and the two main floor inlets. In all cases the power unit should never be mounted more than 8 feet above the lowest inlet; any distance below the lowest inlet (within reason) is fine. A central location of the power unit in relationship to the inlets is also recommended. Figures 21-3, 21-4 and 21-5 show the power unit location in various types of homes.

While the power unit can be located in a closet provided that the door is louvered to provide ventilation, a basement, garage, utility room or carport is better to keep the noise of the power unit away from the living area.

## PLANNING THE TUBING SYSTEM

Once the inlets have been located and the power unit installed the routing of the PVC (plastic) tubing can be laid out.

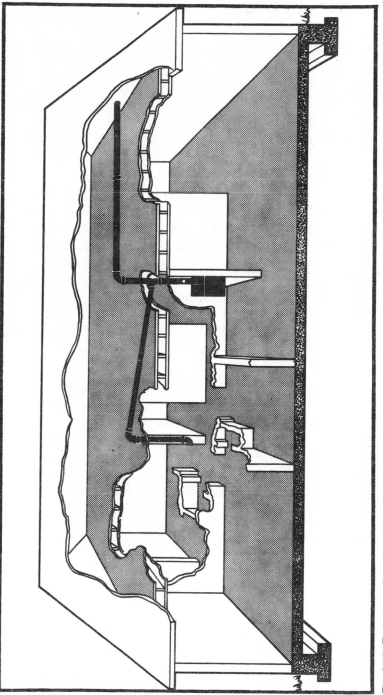

Fig. 21-3. Typical one-floor installation without basement.

This portion of the system is composed of two basic sections: the main trunk line that connects several branches, and branch lines that connect to the inlets as shown in Fig. 21-6. In planning the system adhere to the following suggestions:

■ Keep all runs of tubing as short and direct as possible as unnecessary bends, offsets and fittings reduce operating efficiency

■ Where vertical runs of tubing must carry dirt upward the rise should be limited to 8 feet if possible, and never more than 10 feet

■ Where structure is of the split-level or multi-story types tubing on the top floor can run upward to the attic space or downward through stud walls, or it may be run horizontally between joists when roughing in during new construction

■ Generally, where tubing for two or more inlets on the upper floor goes upward it makes a simpler layout to tie them together in the attic space and then run one vertical drop downward at the most convenient point to the power unit as shown in Fig. 21-6

The planning of existing structures are not quite as easy. It is often necessary to go a roundabout way to get to where you wish to install the inlets. It may be necessary in two-story dwellings, for example, to run the tubing along side a plumbing vent pipe or perhaps alongside an air duct to get to the second floor. Closets are often located one over the other, and in this situation, the tubing could be run up through them in a back corner of each.

## INSTALLING TUBING & FITTINGS

Size tubing used for most residential central cleaning units is 2-inch PVC (plastic) tubing. You should use a slow speed half-inch drill motor with a self-feeding 2½-inch diameter bit to drill the required holes for the tubing. A right-angle drill attachment can save energy, time, and perhaps scratched knuckles. This type of drill motor is especially useful for boring holes in an existing structure.

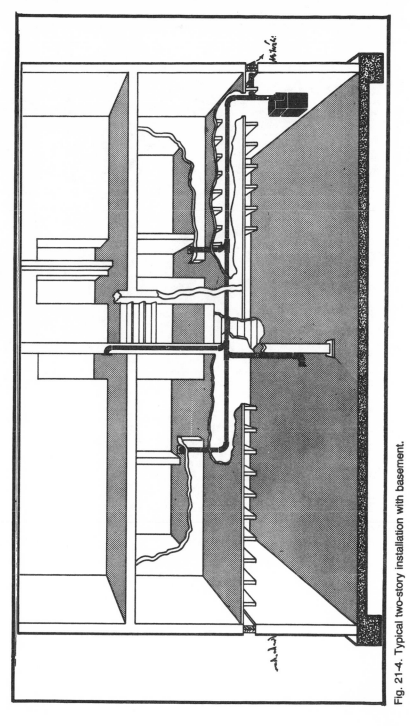

Fig. 21-4. Typical two-story installation with basement.

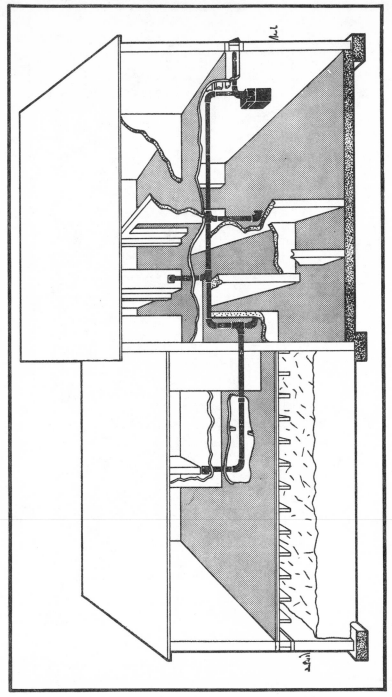

Fig. 21-5. Typical split-level installation.

Vertical risers in existing partitions can usually be run directly in stud walls and fire bridging as shown in Fig. 21-7. Where drilling becomes impractical other convenient locations may prove better and easier. Some of these, as mentioned previously, could include the inside of closets, walls, beside plumbing vents or within cold air return ducts.

When cutting the tubing itself always make the cut as squarely as possible to insure a tight seal between the tubing and the fitting. Either a tubing cutter or conventional hacksaw will do.

Once the tubing is cut ream the inside and remove all the burrs from the outside that could prevent an airtight joint. There are tools made specially for working with PVC tubing, but a regular file will do for deburring and reaming.

After making the cut in the tubing and in order to connect joints together, always apply the cement to the outside of the tubing. Then press the tubing into the coupling or fitting as shown in Fig. 21-8. Never put the cement on the inside of the

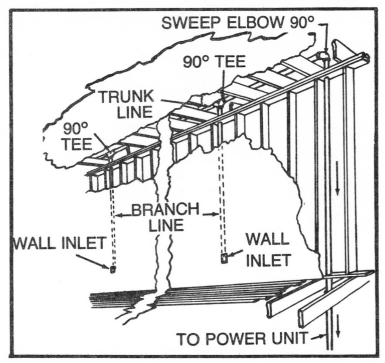

Fig. 21-6. Pictorial drawings showing main trunk line and branches.

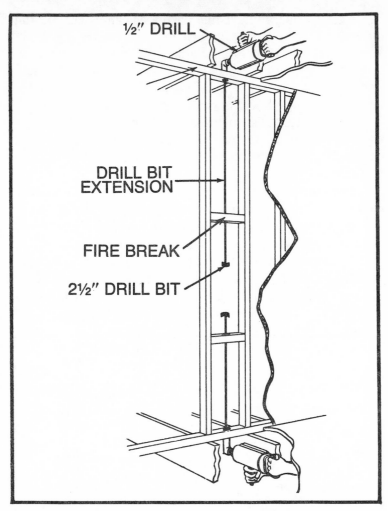

½" DRILL

DRILL BIT
EXTENSION

FIRE BREAK

2½" DRILL BIT

Fig. 21-7. Method of preparing stud walls and fire bridging to accept PVC tubing.

coupling or fitting because when the tubing is inserted the cement will be pushed up to the center and leave a jagged or rough edge of cement. This rough edge will restrict air flow and could lead to stoppage at a later date. The cement sets up rapidly, so don't hesitate once the glue is applied. Should it accidentally get inside the tubing wipe it out immediately.

Several factory-made fittings are available for connections, turns and taps. Figure 21-9 shows two 90-degree elbows, a short and sweep. Use the long radius elbows where

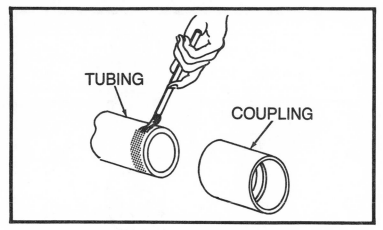

Fig. 21-8. Method of applying cement to PVC tubing.

possible in order to cut down on friction loss in the tubing. However, there will be times when the use of the short elbow becomes necessary, but still try to use it only in the exhaust line.

For taps from the main trunk line, T-fittings (Fig. 21-10) and Y-fittings (Fig. 21-11) are available. In placing or installing these types of fittings always make certain that their position is so arranged to allow the air to flow back toward the power unit.

When the branch tubing enters the main trunk line from opposite directions, a 90-degree elbow and a T-fitting are better than a three-way elbow; both are shown in Fig. 21-12.

Where tubing is run inside a closet rather than in partitions incorporate fittings as shown in Fig. 21-13. You may use

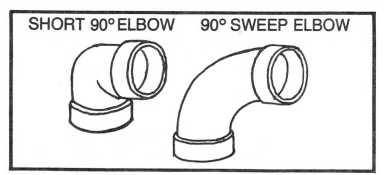

Fig. 21-9. Two PVC elbows; one short and one sweep ell.

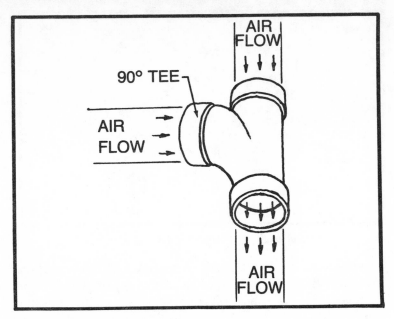

Fig. 21-10. PVC T- fitting.

a feedthrough flanged T-fitting for split-level or multi-storied house as many times the tubing will be stubbed up from underneath the floor as shown in Fig. 21-14. Then use an inlet adapter to place an inlet on the first floor, and couple straight up to the second story floor to an additional inlet.

Although it is important to use as few fittings as possible in order to cut down on air flow resistance certain obstacles will be found that will have to be bypassed and 45-degree

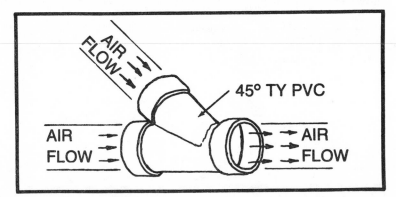

Fig. 21-11. PVC Y-fitting.

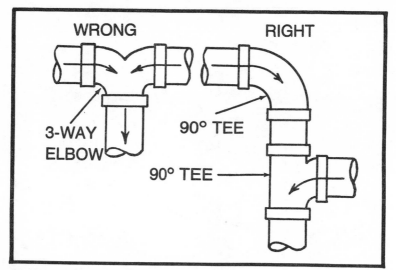

Fig. 21-12. When changing direction in a pipe run, a 90-degree elbow and T-fitting (on right) is better than a three-way elbow (left).

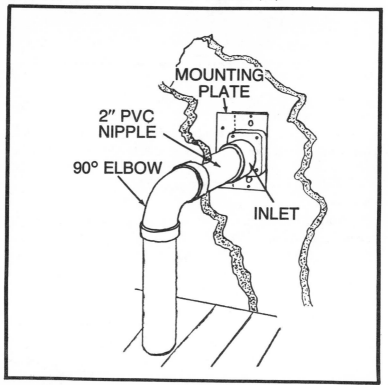

Fig. 21-13. Method of running tubing inside of closet.

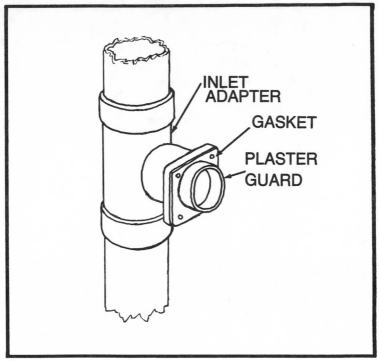

Fig. 21-14. Flange T-fitting.

fittings can be used to run the PVC tubing around the obstruction as shown in Fig. 21-15.

All horizontal runs of tubing should be supported with proper straps not more than 6 feet apart. The tubing may be secured to the bottom of floor joists as shown in Fig. 21-16 and should be anchored at all turns to prevent separation or movement.

When installing the tubing in new construction always use nail guards where the tubing is run through the soleplate or where nails could be driven into the tubing system. This is one thing that will lead to stoppage and cause a lot of difficulty when going back at a later date to locate and remove the nail. Typical nail guards can be seen in Fig. 21-17.

Many times it is possible to run the PVC tubing across the basement alongside air ducts or soil pipe where the ceiling may be furred down, making it not necessary to drill through the joists. It is also possible to run the pipe parallel to the joists to the outside wall or run alongside other piping where the

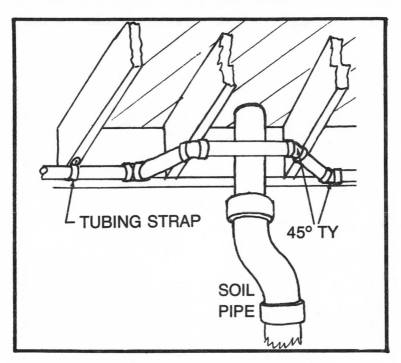

Fig. 21-15. Method of using 45-degree fittings to bypass obstacles.

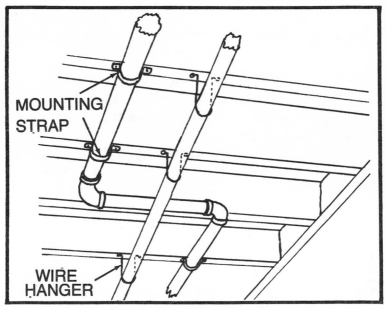

Fig. 21-16. Method of supporting pipe to floor joists.

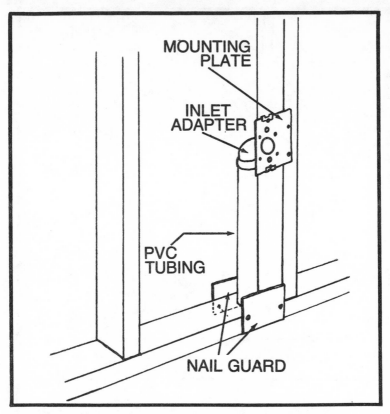

Fig. 21-17. Nail guards used to protect piping in new construction.

ceiling is furred down. However, never run tubing where it will come into contact with steam or hot-water pipes.

It is recommended that most models of central vacuum systems be exhausted to the outside of the building by means of 2-inch PVC tubing along with the standard fittings as shown in Fig. 21-18. Never exhaust into a closet wall stud space. Rather, use an exhaust vent with back draft damper, and exhaust through outside walls.

After the tubing has been roughed in it is very important to test the system at that point (before it is covered) to make sure that you have an airtight system. This can be accomplished by hooking up a power unit and plugging all but one inlet. At that inlet take a sealed vacuum reading to determine how much leakage there is in the system as shown in Fig. 21-19.

276

There usually should be no more than a 5-inch drop (inches of mercury) by taking a reading at the power unit and a reading at an inlet with all other inlets sealed off. A leak found at this stage (before the system is covered) is much easier to locate and repair. Bear in mind that in many new homes there will be a voltage drop due to the temporary service coming to the job site or into the house. This can affect the output of the

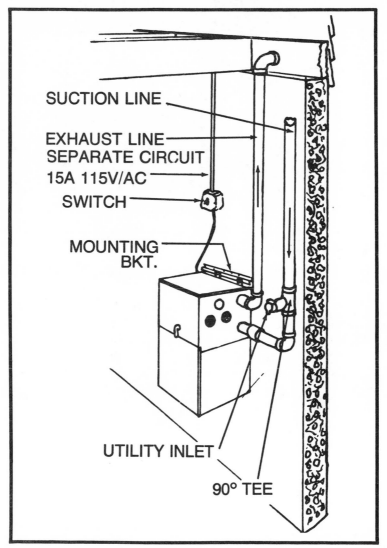

Fig. 21-18. Method of exhausting power unit to outside.

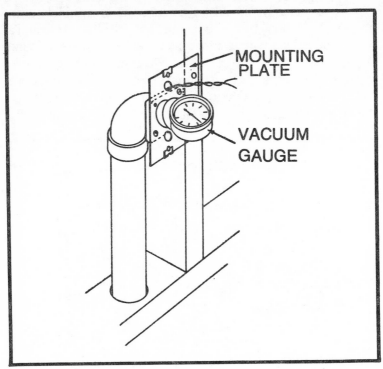

Fig. 21-19. Vacuum gauge used to test piping at inlets.

vacuum motor and is why it is very important to take a reading at the power unit as well as at the inlet. Then you will know how much actual loss you will have in the system. The placement of the vacuum gauge at the power unit is shown in Fig. 21-20.

The installation of the hose inlets would be the next logical step. However, before drilling any holes or cutting wall openings, make certain that your dimensions are correct. Also check that there are no obstructions below the point you are planning to come through, such as a heating duct, plumbing pipes or electrical wires. This can easily be done by driving a finishing nail or drilling a small hole beside the molding opposite the proposed hole location. The nail or drill can be located underneath the floor to make sure there are no obstructions to prevent drilling. Figure 21-21 illustrates this method.

When installing the system in an existing installation, locate the center line of the inlet approximately 14 inches above the finished floor. With a fine-toothed keyhole saw cut an opening 2½ inches wide by 4¼ inches. Most inlet flanges have very little overlap to cover errors, so make certain that your measurements are absolutely correct. Next drill a 2½-inch diameter opening directly below the center of the wall opening for the 2-inch PVC tubing. All of these procedures can be seen in Fig. 21-21.

Wiring for the low-voltage controls should be looped from one inlet to the next before installing the tubing or wall inlets. Run the wiring through the same openings drilled for the tubing and leave approximately 12 inches of wire at each inlet opening. Secure this wire at the opening to prevent accidental withdrawal prior to completion.

The tubing comes next and should be inserted so that the top of the 2-inch PVC extends 1 13/16-inches above the lower edge of the rectangular wall opening as shown in Fig. 21-22. Apply cement to the outside end of the vertical 2-inch tubing,

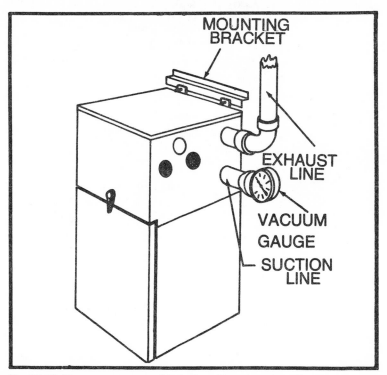

Fig. 21-20. Vacuum gauge used to test system at power unit.

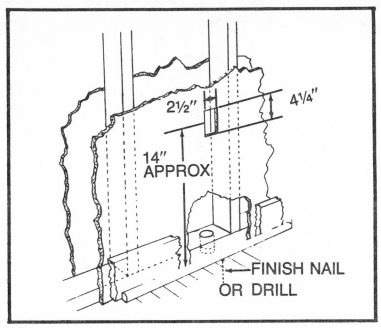

Fig. 21-21. Measurements for roughing in system in existing construction.

and insert the inlet adapter through the opening. Push it securely onto the 2-inch PVC tubing, referring to Fig. 21-22. Thread the low-voltage cable through one of the holes in the mounting plate (Fig. 21-23), and insert this plate through the wall opening. The mounting plate is then secured with the two mounting screws as shown in Fig. 21-24. Be sure to install the gasket between the inlet adapter and the mounting plate. If the wall is less than a half-inch thick it will be necessary to inset shims as illustrated in Fig. 21-25.

Once the mounting plate is in place connect the low-voltage wires to the inlet plate with two wire nuts as shown in Fig. 21-26. Remove the plaster guard (Fig. 21-25) and insert the neck of the inlet plate onto the adapter. Then push it into place. A hooked wire or rod inserted through the opening of the inlet will help to hold the adapter while forcing the inlet into position. The inlet plate is then secured with two screws.

When installing the system in new construction where the walls are open the procedure is the same except that you will have less cutting and fishing. After locating the opening drill a 2½-inch diameter hole in the center of the stud plate with the center of the hole 1 9/16-inches from the wall stud (see Fig. 21-27). Nail the mounting plate to the wall stud approximately 14 inches above the floor and with the inlet opening directly above the hole in the stud plate. Attach the inlet adapter to the wall plate with two 10-32 screws. Then run the PVC tubing from the inlet adapter back to the point where it connects to the main feeder line.

The low-voltage wiring can be taped or wrapped around the 2-inch PVC tubing, brought through the opening in the bracket and a knot tied in it so that it cannot be pulled back

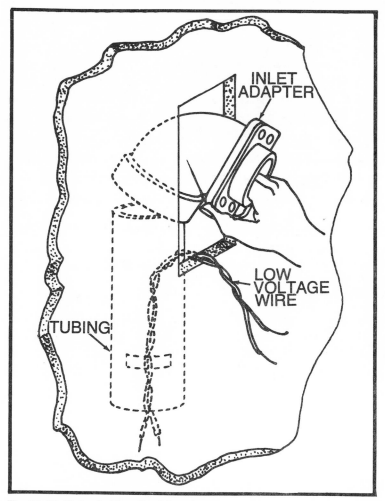

Fig. 21-22. Method of inserting inlet adapter.

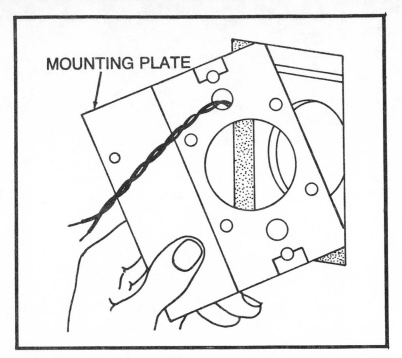

Fig. 21-23. Threading the low-voltage cable through the mounting plate.

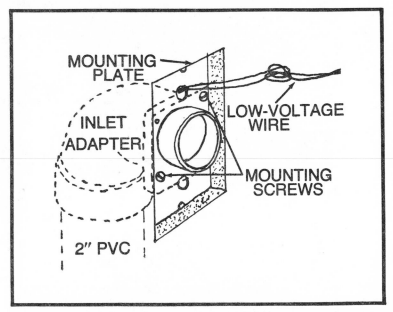

Fig. 21-24. Method of securing the mounting plate with two screws.

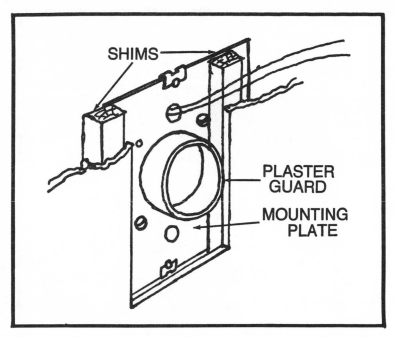

Fig. 21-25. Shims must be inserted if the wall is less than a half-inch thick.

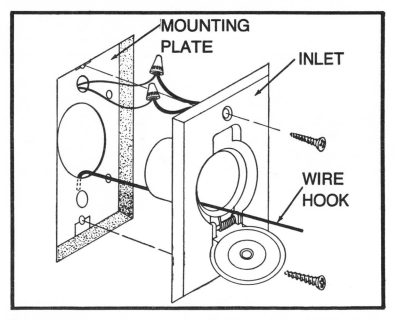

Fig. 21-26. The low-voltage wires are connected to the inlet plate with two wire nuts.

283

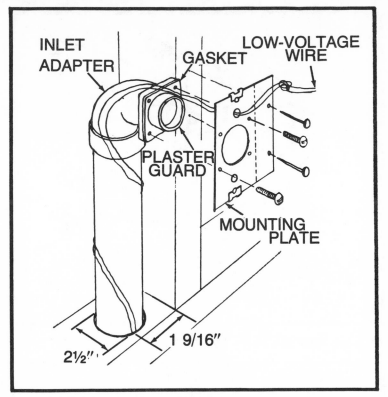

INLET ADAPTER

GASKET

LOW-VOLTAGE WIRE

PLASTER GUARD

MOUNTING PLATE

1 9/16"

2½"

Fig. 21-27. Measurements for roughing in piping for new construction.

accidentally. Leave approximately 12 inches of wire at the opening. The remaining steps are the same as described for installing the system in existing structures.

The installation of the power unit comes next. For existing construction, Figs. 21-28, 21-29 and 21-30 give the steps involved, while Figs. 21-31, 21-32 and 21-33 give the procedures for installing the unit in new construction as well as the connections for both types.

## FINAL CHECKOUT

When the entire system is in place, a final check and test should be made and will consist of the following:

- Make sure that all inlets are closed and the filter bag and secondary filter are in place before starting the vacuum motor

284

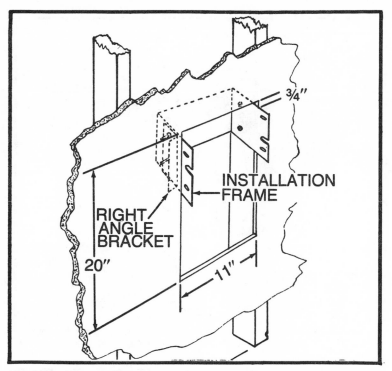

Fig. 21-28. Measurements for roughing in mounting bracket for power unit for existing construction.

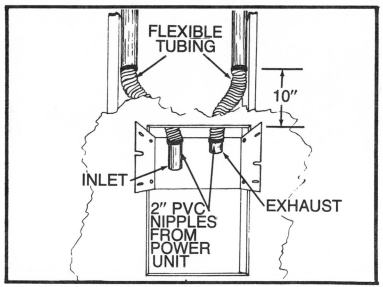

Fig. 21-29. Method of installing flexible tubing to the power unit.

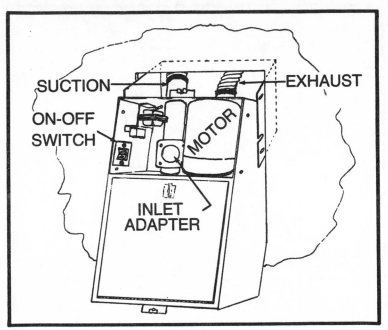

Fig. 21-30. Further details of installing the suction and exhaust lines to the power units.

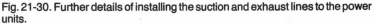

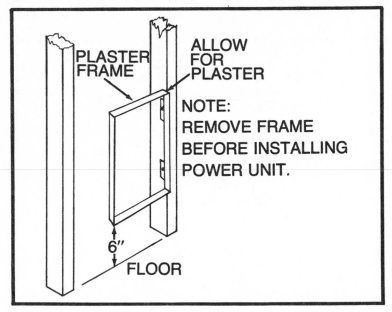

Fig. 21-31. Method of roughing in mounting bracket for the power unit in new construction.

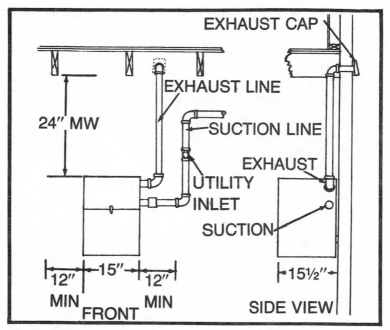

Fig. 21-32. Dimensions for installing PVC tubing to the power unit in either old or new construction.

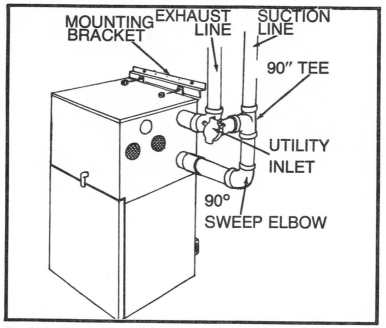

Fig. 21-33. Detail of pipe connections for either old or new construction.

- Open one inlet at a time. Insert metal end of cleaning hose into inlet. Vacuum motor will start immediately and automatically. Remove hose and the unit will cut off automatically.
- Check the vacuum. Insert hose again. Place your fingers over the cleaning end and you should feel a violent rush of air through the fingers. If this fails to occur there is probably an open joint somewhere in the tubing or else a malfunction at some other point.
- Use a vacuum gauge if a leak is suspected and measure inches of mercury at the inlet. Then separate the suction tube at the power unit to check the vacuum at this point.
- Recheck vacuum drop. Replace vacuum suction tube at power unit. If all checks out the system will be in proper working order.

# Chapter 22
# Installing Add-On Air Conditioning

The inclusion of central air conditioning in new homes was considered a prohibitive luxury for many homeowners only a few years ago. They had to be content with only a heating system and perhaps a small window air conditioner in the bedroom or family room. But, central air conditioning now comes in kit form with costs cut to a bare minimum. The average handyman can make the installation in a six-to-eight room house that has forced-air heat with adequate blower and ductwork for under twelve hundred dollars, perhaps even less if the components are purchased in the fall or winter during the off season.

The working principle of this system is simple as can be seen in Fig. 22-1. A compact cooling coil is installed in the furnace plenum or warm-air outlet in the furnace. Then refrigerant lines are run from the coil to a remote condensing unit mounted outdoors. When the power and control wiring is connected and the unit energized the existing fan or blower in the furnace pushes air through the coil and into the existing ducting to every room in the home. In winter the furnace runs as usual as the condensing unit is turned off.

There are several companies manufacturing central residential air-conditioning components designed specially for the do-it-yourselfer. Most of these kits require very little skill and only a few hand tools, all found around the home or farm, and are necessary to complete the average installation in a single day.

Costs of the kits begin around seven hundred dollars for a 2-ton unit (24,000 Btuh) to over one thousand dollars for a 4-ton (48,000 Btuh) unit. The length of the refrigerant tubing, the type of controls and the possibility of needing a replacement blower for your furnace—in case your present one is too small—are other factor which will affect the costs of the kits. Add seventy-five dollars for a concrete pad and another one hundred dollars for miscellaneous materials and we come up with a total cost of between one thousand and fifteen hundred dollars for the complete installation.

In general, there are two basic steps to adding central air conditioning to your existing forced-air system, an analysis of your home's cooling requirements and selecting and installing the components.

This chapter is designed to give you a general knowledge of the installation procedures. Then, if you're convinced that you want to tackle the job yourself exact step-by-step instructions can be obtained when you have chosen the type of kit you intend to purchase or when you pick up the components.

A careful analysis of your home's cooling requirements is an all-important step to help you select the proper size and type of equipment needed for your particular home and existing forced-air system. It is important because an undersized unit will not provide enough cooling to do the job and an oversized unit not only costs more money, but can also cause as much dissatisfaction as an undersized unit in terms of comfort. An oversized unit will be cutting on and off all day with long periods of not running; such an operation will not maintain an even temperature nor will it dehumidify properly.

In most areas an engineer at your local utility company will be happy to make the analysis for you, calculating your exact cooling requirements along with the yearly operating costs. This is often done at little or no charge to the homeowner. If you want to do the job yourself, *Do-it-Yourselfer's Guide to Modern Energy-Efficient Heating & Cooling Systems* (#903), published by TAB BOOKS, has a

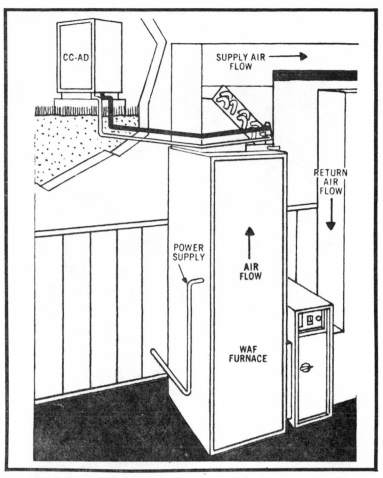

Fig. 22-1. Basic components of an add-on cooling system.

chapter describing in detail how to make cooling calculations
for the home.

　　With the calculations complete your dealer will be able to
recommend the correct cooling equipment for your system,
quote an exact cost and furnish you with detailed instructions
for installing the equipment in your home's forced-air heating
system. Your kit will look something like the one illustrated in
Fig. 22-2 and will consist of two lengths of refrigerant tubing,
four baffles, plenum cover, coil supports, sealers, cooling coil,
condenser-compressor unit and complete step-by-step instructions so that you can't go wrong.

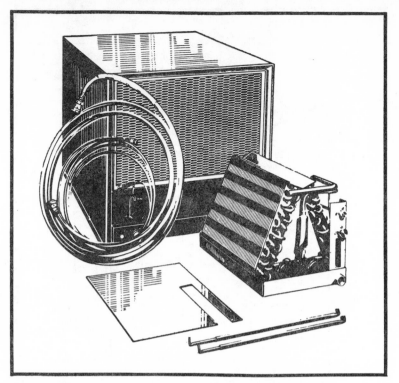

Fig. 22-2. Typical add-on cooling kit for a forced-air system.

## MOUNTING THE CONDENSING UNIT

Prior to ordering your kit, select the exact spot you intend to mount the outside section (condensing unit). This will enable you to order the correct length of refrigerant tubing.

All air-cooled condensing units must be mounted outside of the building, as near as possible to the cooling coil inside of the furnace plenum and on solid level supports such as those shown in Fig. 22-3. Proper clearance must also be provided for the air intake and the air discharge, refrigerant piping and power connections. Clearance, of course, must also be provided for servicing and maintenance of the unit.

## INSTALLING THE COOLING COIL

Most cooling coils designed for the do-it-yourselfer are charged with a specific holding charge of refrigerant and sealed at the factory to help simplify the installation. You will need only a few hand tools for this job, that is, tin snips, electric drill, screwdriver, knife, hammer, scratch awl, level, drop light and ⅛-inch drill bit.

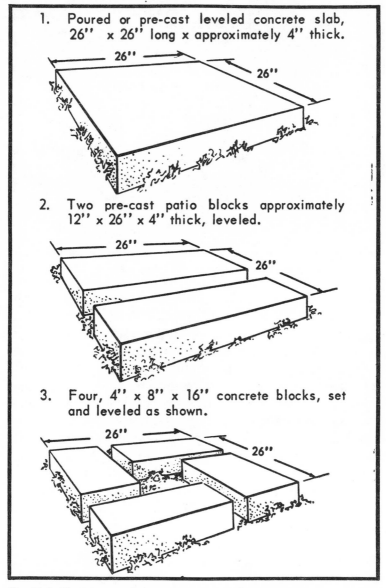

1. Poured or pre-cast leveled concrete slab, 26″ x 26″ long x approximately 4″ thick.

2. Two pre-cast patio blocks approximately 12″ x 26″ x 4″ thick, leveled.

3. Four, 4″ x 8″ x 16″ concrete blocks, set and leveled as shown.

Fig. 22-3. Three types of supports on which the condensing unit may be mounted.

Fig. 22-4. Cutting the opening in the furnace plenum with tin snips.

You begin with starting a hole in the furnace plenum by placing a screwdriver at an angle against the sheet metal, then strike a sharp blow with a hammer. With tin snips, cut a small opening to enable you to examine the interior of the plenum to ascertain height and position of the inside duct flanges. Once the flanges have been located continue to cut out the opening (Fig. 22-4) to dimensions slightly larger than the height and width of the cooling coil.

Your next step will be to place the support bars in place, then cut the baffles to fit. With these in place a sealing compound is used to insure that all air goes through the cooling coil. All of these components can be viewed in Fig. 22-5. Distance A in Fig. 22-5 must be less than the width of the coil assembly.

You now have the hardest parts completed and you're more than halfway finished. So why not take a short break and try to round up a neighbor to help you lift the cooling coil in place.

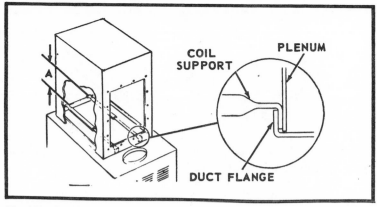

Fig. 22-5. Placement of the coil supports.

To place the cooling coil inside of the plenum, grasp the coil by the tubing at point A as shown in Fig. 22-6, and the condensate tray at point B. Always keep the top of the coil near the top of the plenum opening while inserting it, and never allow the bottom of the coil to move the seal out of its proper position. If it does you will have to reseal the baffles with caulking compound.

When the cooling coil is in place and you're certain that an airtight seal has been accomplished replace the opening cover, insert and tighten all screws and you're just about ready to connect the refrigerant lines. But first install a ¾-inch

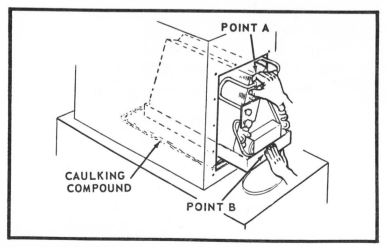

Fig. 22-6. Method of inserting the coil into a plenum.

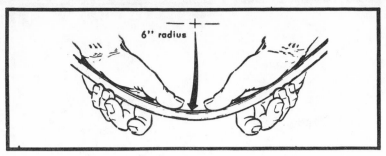

Fig. 22-7. Smaller tubing can be formed by hand.

condensate line from the condensate tray to the closest floor drain, sump pit or similar place where the condensate liquid can be properly disposed of.

Fig. 22-8. Larger tubing is formed with a spring bender covering the area to be bent, then placed over the knee. Care should be taken not to kink the tubing.

## INSTALLING THE REFRIGERANT TUBING

Since the cooling coil is usually inside of the house and the condensing unit outside holes of the proper size will have to be

Fig. 22-9. Tightening refrigerant line couplings with wrenches in the correct manner.

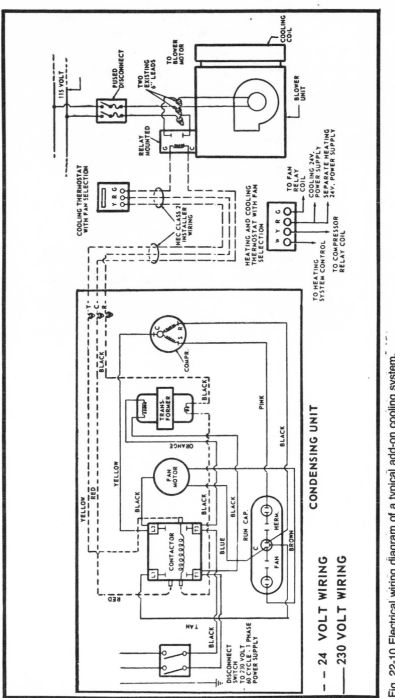

Fig. 22-10 Electrical wiring diagram of a typical add-on cooling system.

cut to allow the refrigerant lines to pass. When running the tubing you will have to do some bending, but since the copper tubing has been annealed it can be formed to look neat with just the hands. The bends of the smaller tubing are made over the thumbs as shown in Fig. 22-7 while the larger tubing is bent over the knee (Fig. 22-8) after inserting a coiled spring tube bender, usually furnished with the kits, to avoid kinking.

The couplings are then connected to their proper places on the coil and condenser, only hand tight at first. After all four connections have been loosely made tighten the couplings, using appropriate wrenches as shown in Fig. 22-9 until a definite resistance is felt. Then tighten an additional quarter-turn.

## INSTALLING THE ELECTRICAL WIRING

Different manufacturers will have slightly different wiring connections, but the wiring details in Fig. 22-10 show the basic procedures.

In nearly every case the condensing unit for a home central air-conditioning system will be rated at 240 volts, single phase. So review Chapter 9 before starting the installation. Chapter 8 will also help you with the low-voltage control wiring. Just remember that for both the 240-volt circuit and the low-voltage wiring all work must meet the requirments set forth in the National Electrical Code and all local ordinances.

With the wiring out of the way you have nothing left to do but follow the manufacturer's instructions on the start-up procedures, then set back and enjoy your central air-conditioning system for years to come.

# Chapter 23

# Electronic Garage Door Controls

Like most Americans you probably have several of the modern conveniences in your home. Your home is probably heated with automatic controls, you can turn your TV set on from your favorite chair and talk from one room to another without taking a step or raising your voice. But how do you tackle the largest single moving part of your house—your garage door? If you still operate the garage door by hand you're missing a lot of convenience and security that you can easily have by installing an automatic garage door system.

That's right. You operate the door by remote control. Day or night, whatever the weather, you'll be dry and comfortable because you stay in the car until you're inside the garage—plus no bending or lifting.

## WHY HAVE ONE?

Let's assume that your wife has gone over to a friend's house for a game of bridge with the girls. When she returns home, it's after midnight, raining like cats and dogs and you forgot to leave the outside lighting on for her to see. With a manual garage door, she would have to get out of the car in the dark, get soaking wet in the rain, fumble for the key, then manually raise the door. Besides getting wet, she would also be subject to attack from prowlers. You could have left the garage door open for her before you turned out the lights and went to bed, but an open garage door advertises that no one is home. It's an open invitation to prowlers or intruders.

On the other hand if your home was equipped with an automatic garage door system, your wife would merely press a transmitter in her car as she pulled up into the driveway. As the garage door automatically opens, a light inside of the garage automatically turns on at the same time, allowing her to see inside of the garage before she enters.

She then drives in the garage but stays in her car until the garage door closes and locks behind her. The automatic light, however, stays on until she has time to enter the house or turn on an auxiliary light. She never needs to be in the dark to stumble over something. The closed garage door is also provided with an automatic lock to help deter even the most aggressive unwanted caller.

If the garage door meets an obstruction while opening or closing, it automatically reverses or comes to a full stop, depending upon the type you install. This helps to protect children, pets and possessions from being trapped under the door.

Regardless of the type of garage door you now have or plan to install there are automatic garage door systems designed to operate them. Most types specially engineered for residential use, however, are designed for use on 7-foot or 8-foot upward-acting doors, either single or double sectional doors and one-piece doors with track or jam hardware.

## INSTALLATION

Figure 23-1 shows a top and side view of an installation detail for a sectional door. In this type of installation, the motor assembly and chain-drive track attaches to ceiling or wood support with angle straps. It should be located in the exact center of your garage door opening with one end of the track attached to the header above the garage door opening using the mounting bracket that comes with each kit. The door-connecting arm that is shipped assembled to the track rail is adjustable for door height clearance. Merely loosen the two bolts, adjust the mounting bracket section to the proper position and secure it to the top of your garage door with the

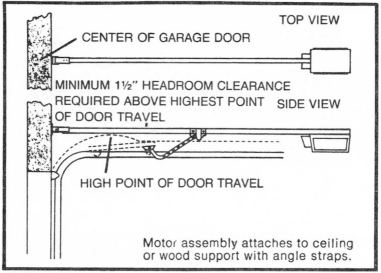

CENTER OF GARAGE DOOR

MINIMUM 1½" HEADROOM CLEARANCE
REQUIRED ABOVE HIGHEST POINT   SIDE VIEW
OF DOOR TRAVEL

HIGH POINT OF DOOR TRAVEL

Motor assembly attaches to ceiling
or wood support with angle straps.

Fig. 23-1. Top and side view of a garage door opener installed on a sectional door.

two or three woods screws that come with the outfit. Run a 120-volt power line from your panel box, or nearby lighting or convenience outlet (as described for duplex receptacle wiring earlier in this book). You're ready for operation once you connect the receiver with only two wires to the 24-volt control circuit of the garage door operator. You then press your remote transmitter to operate the garage door.

A typical installation would begin by ordering a door system designed for your particular type of door, that is, upward-acting single or double sectional doors or a one-piece door with track or jamb hardware. A sectional door installation is shown in Fig. 23-1, a one-piece door installation is shown in Fig. 23-2 and a one-piece door with jamb hardware is shown in Fig. 21-3. In all situations, make certain that you have enough head room for the track before ordering the system.

Since the one-piece door with jamb hardware is the most complicated to install, we will use this type as our installation example.

Before beginning any mounting of the automatic door system, operate the garage manually several times in order to determine the lowest and highest points of the door top as shown by the dotted line in Fig. 23-3. When you are certain

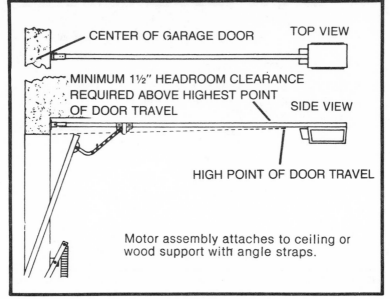

CENTER OF GARAGE DOOR    TOP VIEW

MINIMUM 1½" HEADROOM CLEARANCE
REQUIRED ABOVE HIGHEST POINT
OF DOOR TRAVEL    SIDE VIEW

HIGH POINT OF DOOR TRAVEL

Motor assembly attaches to ceiling or
wood support with angle straps.

Fig. 23-2. Top and side view of a garage door opener installed on a one-piece door.

that you have correctly determined these points clearly mark each one.

Next line up the motor housing in the exact center of the garage door and level with the lowest point of the open door. Position it only, do not secure firmly. Elevate the track or T-rail end that is opposite the motor housing to an angle required to clear the highest point of the door's path of travel. Using the screw holes in the mounting bracket as a guide mark the location of the screw holes on either the ceiling or wall, whichever the bracket is to be mounted too. In choosing this location make certain that the structure is strong enough to mount the bracket.

If the structure to which the bracket is attached is wood conventional wood screws supplied with the kit will suffice. However, if the attaching point is masonry, toggle bolts will have to be used for attaching to hollow concrete block, and lag or other type of masonry anchor will have to be used in brick or solid masonry walls or ceilings. For the former (toggle bolts) drill a hole through the masonry to the hollow area, large enough for the bolt and wings to pass. In solid masonry drill a hole with either a masonry drill or star drill and hammer. The size of the hole should be just large enough to accept the lead sleeve and deep enough so that it fits flush with the wall or ceiling surface prior to setting the anchor. A special tool for setting the anchors is sometimes supplied with the anchors. The method of installation is shown in Fig. 23-4.

Once the door end of the track or T-rail is secure, mount the motor housing in its proper position (level with the lowest point of the door top as it travels). Where the ceiling consists of wood joists the motor housing can be secured directly to the sides of the joists with wood screws. In masonry walls or ceilings a special bracket will have to be secured to the wall or ceiling first with lead anchors or toggle bolts before the motor housing can be attached to it with bolts and nuts.

The next logical step would be to mount the connecting arm from the track of the T-bar to the door. Mount this arm directly in the center of the garage door and near the top of it. Wood screws are provided with the kit for installing the arm to the garage door, and bolts and nuts are used to connect the arm to the track of the T-bar. For a one-piece door with jamb

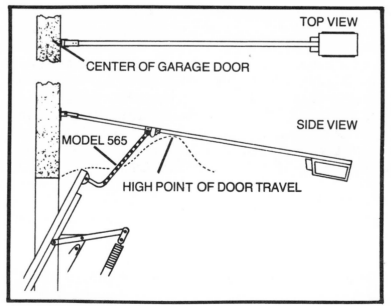

Fig. 23-3. Top and side view of a garage door opener attached to a one-piece door with jamb hardware.

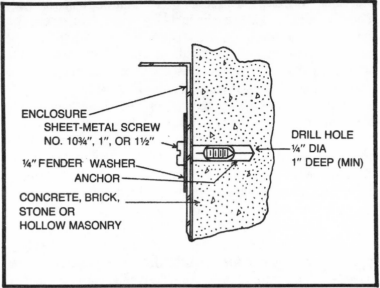

ENCLOSURE

SHEET-METAL SCREW
NO. 10¾", 1", OR 1½"

¼" FENDER WASHER

ANCHOR

CONCRETE, BRICK,
STONE OR
HOLLOW MASONRY

DRILL HOLE
¼" DIA
1" DEEP (MIN)

Fig. 23-4. Method of securing items to masonry walls. First drill the hole with a star drill or masonry drill bit. Then insert the lead or plastic anchor. The anchor should be snug in order for it to hold. Finally with the item in place insert the sheet-metal screw or lag bolt.

hardware, a longer-than-usual arm, about 4 feet long, designed to operate this type of door more smoothly and efficiently should be used.

Drive motors used on most residential garage door systems usually are rated between one-quarter to one-third horsepower, each pulling approximately 5 amperes of current. Therefore, AWG #14 will handle this load quite well. In fact if the circuit feeding the garage lights is not fully loaded it is quite permissible to feed the drive motor of the operating system directly from one of the nearby lighting outlet boxes.

## OPTIONAL ACCESSORIES

You may also want to include—out of necessity or convenience—one or more of the following optional accessories.

**outside key-operated manual release**—Allows manual release of operator from outside garage where a secondary entry is not provided. It mounts on face of garage door as close as possible to operator release mechanism.

**low headroom kit**—for use with sectional garage door installation only without sufficient headroom clearance, that is, less than 1½ inches for operator.

**installation strapping**—Galvanized angle strapping with 11/32-inch holes on 1-inch centers for securing garage door operator to ceiling joists.

# Chapter 24
# Major Appliance Considerations

Frequent additions to existing residential or farm electrical wiring systems are appliances of one sort or another, or the installation of new equipment or machinery. These added loads may not, but often do, necessitate the running of new circuits or the installation of a new service entrance for increased capacity and may even instigate a rewiring job.

If rewiring is required or new circuits must be run or the service-entrance equipment has to be changed refer to the applicable information in earlier chapters. The following pages will give you detailed information on installing various specific units useful for making allowances for future additions, for installations in a new system or for additions to an old one.

## COMPACTOR

The trash compactor is becoming an increasingly popular appliance and is one of the easiest to install. This appliance is designed to operate on 120-volt circuits and is equipped with a short cord and attachment plug assembly for quick connection to a convenience receptacle. The plug is three-pronged and should be used with a grounding receptacle so that the metal frame of the compactor is adequately grounded. If grounding proves to be impractical you may be able to omit it, depending upon local codes.

Most compactors draw between 400 and 500 watts. As they are motor driven there is a certain amount of starting current surge. There is also the possibility of their working under heavy load or jamming, which means high potential current draw. On the other hand, they are used infrequently and only operate for a short period of time. You could run a separate circuit to a compactor, either 15 or 20 amperes. You could also wire it to an existing circuit that is not loaded to more than about half of its 80 percent capacity, provided that no lighting fixtures are on the same circuit.

## DISPOSAL

The garbage disposal is a fixed appliance and installation is governed by local codes in many places. Made to operate on a 120-volt branch circuit, most consume about 400 to 500 watts of power, but some may go as high as 1000 watts. A disposal is permanently installed in a drain line under the sink, and it usually is wired directly from a wall junction box by means of suitable tap conductors run in a length of flexible conduit to a connection box within the unit. Most are operated by a wall switch located near the sink. The disconnecting means can be that same controlling switch, backed up by the main switch. Or, it can be the branch circuit breaker if readily accessible.

The preferable means of installation is to provide a short cord and three-pronged attachment plug to connect to a grounded receptacle located close to the disposal. This method is preferred because a disposal is sometimes installed and connected by a plumber, rather than electrician, who may not know the value of the ground connection. Or, the unit may be removed by a serviceman with the same lack of knowledge and replaced without benefit of ground. With a grounding plug attachment, there is no chance for mistakes.

The disposal is motor driven and has a start-up current surge plus the possibility of high current draw if it bogs down or jams. Most are protected internally against burnout or overheating of the motor with a resettable protective device. They should be, and are often required to be, supplied by a separate 20-ampere circuit and wired with AWG #12.

## REFRIGERATOR & FREEZER

Domestic refrigerators and freezers are made to operate on 120 volts and are supplied with cords and attachment plugs. A few might be considered as fixed appliances but most are stationary, and the disconnecting means is the attachment plug and receptacle combination. Most units consume only a couple hundred watts of power, though some may go as high as 400 watts. The usual procedure is to plug them in anywhere, without much regard for the rest of the circuit load. But they are motor driven, thus have a starting surge, so plenty of margin is desirable.

Separate circuits are a good idea for both appliances to insure greater safety for their expensive contents if nothing else. If another load on the refrigerator circuit trips the circuit out unnoticed, the result could be a major household crisis. The separate circuits can be either 15 or 20 amperes. If either appliance is placed on a circuit with other loads take care that the circuit is only lightly loaded and has no chance of being used for portable appliances of high wattage.

Most appliances that operate on 120 volts are not required under the NEC to have their frames grounded, but refrigerators and freezers in household use are two important exceptions. Both must be adequately grounded. With new equipment and a fairly new electrical system that includes a grounding equipment conductor and three-slot receptacles, this is no problem. But older units will have two-prong plugs and two-conductor cords, and the existing receptacles may well be nongrounding. In this case you should convert them if possible, and if you cannot, it may be necessary to provide a separate grounding conductor from a nearby cold-water pipe to the appliance frame.

## CLOTHES WASHER

Clothes washers are made to operate on a 120-volt branch circuit and are supplied with a cord and attachment plug rather than being directly wired. This is another of the excepted instances when the appliance must be adequately grounded. If you have an ungrounded electrical system examine the possibilities of installing a new grounded circuit, or a grounded extension to an existing circuit, using the three-slot grounded receptacles.

Small washers may use as little as 60 watts of power. The medium range seems to be about 400 watts, but many large units take as much as 900 watts. This means at the least that you should carefully size the specific washer load to an existing circuit, using either a lightly loaded circuit or one which has loads that are not likely to be used concurrently with the washer. This is a motor load, so the starting surge has to be considered. Large machines should be supplied by a separate circuit, or one that has only a small lighting load, for instance, with no chance of other appliances being plugged in along the line. The disconnecting means is usually considered to be the attachment plug.

## CLOTHES DRYER

There is a wide range of clothes dryers available on today's market. The smallest are made to run on a 120-volt branch circuit and consume about 1400 to 1500 watts of power. These must be fed by an individual 20-ampere branch circuit. Other types are designed for 208/240-volt operation, fed by an individual three-wire branch circuit. Power consumption may run from about 4400 to 7000 watts for automatic dryers, while the high-speed type may consume as much as 9000 watts. The common practice is to provide conductors matched to the size of the present dryer load, but a better practice is to install a heavy line adequate for the largest types, though with the overcurrent protection device sized to the *present* load. Then later replacement with a larger size dryer can be made with no difficulties.

A dryer is a combination load, involving both a motor load and a heating load. To arrive at the proper conductor size, find the current draw of the motor and multiply that by 125 percent. The heating load will probably be given in watts; divide that by the source voltage and add the answer in amperes to the motor amperage.

If your voltage source is somewhat different than listed on the motor you might wish to convert for greater accuracy. For instance, your motor load is 5 amperes at 220 volts, but your source is 208 volts. Find the wattage ($P = IE$); multiply

220 times 5 to obtain 1100 watts. Then divide the wattage by 208 ($I = P/E$) for an answer of 5.3 amperes. Then multiply that by 125 percent for a total of 6.6 amperes motor load. Assume the heat load wattage to be 5000; divide that by your source voltage of 208, for a current draw of 24 amperes. Add the two, and your total load is 30.6 amperes. By consulting Table II-1 (Appendix II) you can see that the nearest conductor size to this figure, in the 60°C class, is AWG #8 copper.

An alternative method of figuring is to take the total wattage of the dryer, convert to current draw, multiply by 125 percent, and pick the appropriate conductor size on that basis. If the total wattage is 8000 and your source voltage is 208, then $I = P/E$ or 8000/208 = 38.46 amperes, times 125 percent equals approximately 48 amperes. From the conductor table, then, the proper 60°C conductor size is AWG #6. Note that in some installations it may be necessary to go to a higher temperature class.

The frames of all dryers must be grounded. This means that a fourth conductor (green or bare) must be included in the cable assembly of the branch circuit. If the circuit is run in a metallic raceway the raceway can be used as the equipment grounding conductor. There is one exception though: if the supplying service is 120/240-volt three-wire or is 120/208 volts derived from a three-phase four-wire supply the neutral and the equipment grounding conductor can be the same as long as that conductor is no smaller than AWG #10 copper. This means that the cable can consist of three conductors instead of four with the neutral (white) also attached to the appliance frame.

An old standby for wiring dryers is type SE cable, which has a stranded bare neutral conductor wrapped around the two hot insulated conductors. There are some fairly new restrictions on the use of this type of cable—it can now be used for this purpose only if the branch circuit originates in the main service equipment, and never from a subpanel or load center. A dryer branch circuit originating from a subpanel must have all insulated conductors, three or four as the situation demands. Also, if your supply source is different than those mentioned above so that the equipment grounding conductor and the neutral cannot be the same the conductors must all be insulated.

A dryer may be wired directly into the system by means of tap conductors, or it may be provided with a heavy-duty molded cord and attachment plug assembly (usually called a pigtail) and a matching receptacle that may be either flush or surface mounted. These come in various standard sizes: 20-, 30-, 40-ampere and so forth. Whenever possible, the plug and receptacle arrangement should be used for ease of service. Every dryer must have a disconnect, which can be the attachment plug if readily accessible, or the branch circuit breaker. It is also permissible to use the main service disconnect or the branch circuit pull-out fuse holder.

## ROOM HEATER

There are dozens of types of electric room heaters, some portable and others built-in. Many operate on 120-volt branch circuits, while others, almost invariably of the built-in type, function on 208/240 volts. The portable variety may consume anywhere from a low of 200 to 300 watts to a high of about 1500 watts. Some models also are equipped with a small fan for air circulation. The portable types should be used with great care and only on individual or lightly loaded circuits, depending upon the wattage of the unit. A 1500-watt unit, for instance, goes beyond the 80 percent loading factor of a 15-ampere circuit; it has only a 3-ampere margin on a 20-ampere circuit. Cord and three-prong grounding plugs are usually provided, and the units should always be well grounded.

The wattage rating of the built-in types may be as low as 500 watts, or range up to 6000 or more. Some are equipped with fans and others are not, and they may be controlled by line-voltage thermostats built into the unit or mounted on the wall, or by low-voltage wall thermostats. Conductors are sized on the basis of 125 percent of the current draw; divide the wattage of the unit by your supply voltage, and multiply by 1.25 to find the answer.

Larger heating units are served by individual branch circuits sized to the load, and the insulation can be in the 60°C class unless a notation to the contrary is stamped on the equipment. They must be properly grounded and have over-current protection sized to the circuit conductors. The disconnecting device can be the branch circuit overcurrent device, or the main disconnect if readily accessible to the user. Several units may be wired to one circuit or controlled by one or more thermostats.

## WATER PUMP

Water pumps are usually installed in a location where they can be served by surface wiring. They may be wired direct to a junction box by means of tap connectors in a length of flex, or they may be equipped with an attachment plug and receptacle. There is a wide variety made to run on 120 volts, more for 208/240 volts, and some are arranged so that they can be wired either way by simply connecting the proper combination of terminals in the connection box. Power consumption runs from about 450 watts on up, depending upon the size of the pump.

Smaller units can be served by a branch circuit that includes additional loads, though the most advisable situation is a separate circuit. A disconnect or service switch may be required at or near the pump, and sometimes overcurrent protection is provided at the equipment in addition to that at the head end of the branch circuit. This protection is sized closely to the normal demands of the motor, so that if it begins to drag or bog down, as when trying to pump against a frozen water line, the circuit will trip out quickly. Thus, a motor with a normal running current of 7.8 amperes might be protected with a time delay fuse rated at 8 amperes, which would trip long before a 20-ampere branch circuit protection device might.

Water pumps must be well grounded. The disconnecting means may be attachment plug and receptacle, branch circuit overcurrent protection device, service switch or a separate disconnect, depending upon local codes, the specific equipment and the installation conditions.

## WASTE AERATOR

This equipment is relatively new to the market and is made under several trade names. Designed to take the place

312

of a standard septic tank, these units break down household waste and sewage by mechanical and bacterial action, and the resulting output is almost completely pure water. They constitute motor loads, and though there are various sizes, none made for residential use draw any great amount of power.

This equipment is made to be buried and so requires an underground branch circuit to supply the power. The usual method is to run AWG #10 or #12 type UF cable, properly buried and nowhere exposed to sunlight. Alternative methods may be required by local codes, such as metallic or perhaps nonmetallic raceway. Installation procedures and methods are the same as for a service lateral. The equipment must be grounded and protected by an overcurrent device sized to the serving branch circuit, with additional protection sized to the motors if desired. Ground-fault current protection would be a good idea.

## ROOM AIR CONDITIONER

Room air conditioners, whether wall- or window-mounted and whether fixed, semiportable or entirely portable, are considered as appliances. Most units designed to operate on 120 volts consume power in the range of 800 to 1500 watts, so they should be supplied by individual branch circuits, though some of the smaller portable types can be plugged into existing general-purpose circuits provided that no other heavy loads are involved. The 208/240-volt models may consume from 2000 to 4500 watts, possibly more, and are served by separate three-wire branch circuits. Two or more can be attached to the same circuit, provided that the loading reaches no more than 80 percent of the conductor ampacity of the circuit.

Most 120-volt types use an attachment plug for connection to the circuit. The 208/240-volt types can be wired direct, but often they too are fitted with attachment plugs for ease of service and removal. The receptacles are similar to those used for dryers and may be either flush or surface mounted. The branch circuit rating should be not less than 125 percent of the current draw of the unit, and the overcurrent protection device can be sized to the capacity of the conductors. Proper grounding is required, and the disconnecting means is generally the attachment plug or the branch circuit overcurrent device, or it can be the main disconnect.

## DISHWASHER

Automatic dishwashers are made in two types: portable, which roll around on castors, and fixed, for undercounter permanent installation. All run on 120 volts and most consume about 750 watts of power, though a few go higher. For the roll-around type, a specific convenience receptacle close to the sink should be provided where the machine can be plugged in. The branch circuit should really be an individual one, or at the least one that will have no other load attached at the same time the dishwasher is operating. The same is true of the built-in models, except that in many localities this type is required to be on a separate circuit.

Fixed dishwashers are generally supplied with only a connection box for direct wiring into the system with tap conductors run in a length of flex to a junction box usually placed in or on the wall behind the machine. A better system is to provide a short cord and attachment plug and a grounded receptacle. This makes servicing easier and avoids the possibility that some serviceman might not replace the grounding conductor properly. The branch circuit may be rated 15 or 20 amperes, with the overcurrent protection device or the main disconnect serving as the disconnecting means. A separate service disconnect may also be required. Proper grounding is an absolute must.

## SUMP PUMP

Sump pumps are generally used in the home to keep the basement clear of water during the spring flooding. All are designed for 120-volt operation and are usually equipped with an attachment plug and cord to be plugged into any handy receptacle.

Smaller models, which use about 500 watts of power, may be plugged into an existing circuit. But since this is a motor load, which as a rule will turn on automatically upon demand, the serving circuit should not be a heavily loaded one, probably not more than 50 percent of current capacity.

Manual types of pumps can be run with other loads on the same circuit and turned off when necessary. Large sump pumps will use as much as 1000 watts of power, and if automatic should be served by an individual branch circuit.

These pumps can also clog easily under some circumstances. If this looks like a possibility, install a time-lag motor fuse assembly in a junction box at some point between the pump and the branch circuit outlet with the fuse sized closely to the normal running current of the motor.

The plug attachment and receptacle can serve as the disconnecting means, but if the pump is permanently installed you may be required to have a motor-rated service switch or other separate disconnect close to the equipment. Proper grounding is particularly important.

## BARBECUE & SMOKER

Electric barbecues and smokers have become increasingly popular over the past few years. Often as not, they have been improperly installed. These portable appliances are designed to operate on 120 volts and are supplied with a grounding attachment plug and cord. Power consumption may run anywhere from 1000 watts to 1500 watts, which means that they should be served by individual branch circuits. Since these appliances are invariably used outdoors that circuit is usually placed underground in a rigid metallic conduit raceway or protective stub, and should terminate in a weatherproof outlet box equipped with a weatherproof grounding receptacle.

The installation of the circuit itself, whether done with direct-burial cable or in buried raceway should be treated as though it were a service lateral.

## VENT FAN

Ventilation and exhaust fans of one sort or another are usually found in the home, perhaps as part of the heating system, or in bathrooms, range hoods, roof ventilators, photographic darkroom air changers, etc. Most are designed to be wired directly to the system, except for the portable plug-in cooling fans, and most draw only a small amount of current. Even large attic venting fans usually draw no more than 3 or 4 amperes. This means that almost invariably they may be connected to any branch circuit, provided the total draw including the fan does not load the circuit to more than 80 percent of its conductor current ratings.

Permanently installed fans should be grounded in the usual manner wherever possible. In some cases, such as a roof or attic venting system where the ambient temperatures are likely to be above normal, insulation of a high-temperature class must be used on the conductors. Usually a minimum 75°C class is adequate. Switch or controller loops should be of the same temperature class. When figuring loads use 125 percent of the nameplate current rating.

## FOOD CENTER

Food processing centers are made to be permanently installed in a convenient countertop location in the kitchen. They are used for blending, grinding, and chopping operations of relatively short duration. They operate on 120 volts and consume around 400 watts of power. As motorized loads, they should be figured at 125 percent of the nameplate rating for total load current draw. They can be readily wired to an existing small appliance circuit or general-purpose circuit that has sufficient load capacity, particularly one that has loads attached that are not likely to be in use at the same time as the food center.

## WALL TOASTER

This is a relatively new appliance that is modeled after the standard kitchen toaster, but is permanently installed in a case recessed into the wall and pulls out on a slide for use. Wiring is done by direct connection to the circuit. The operating voltage is 120 volts, and the power consumption may run from 800 watts to 1600 watts. A large unit should be served by an individual 20-ampere circuit; a smaller one can be connected to a small appliance circuit that has sufficient remaining load capacity, or where there is ample diversification of existing loads so that there is no chance of overload. Proper grounding is essential.

## FIRELOGS

Electric fireplaces and firelogs are available in two different types: those that produce heat and those that do not. The nonheating variety only draws 60 to 100 watts of power, about the same as a light bulb, and so can be attached to almost any circuit. Such models are equipped with a cord and attachment plug. There seems to be a rule that says there is never a receptacle right where you want to position the fireplace, though, so you will probably have to arrange a branch circuit extension.

The heating type of electric fireplace is also made to operate on 120 volts, and it can be connected by means of a special high-temperature heating appliance cord and attachment plug, or wired direct. Power consumption is about 1500 watts. This may be considered as a continuous load since it could easily be operated for more than 3 hours at a time. For this reason the supplying branch circuit should be AWG #10 because the 80 percent loading factor is exceeded on a 20-ampere circuit. This would hold true for any unit where 125 percent of the rated current, computed on the basis of your own voltage supply, would exceed 16 amperes. A 30-ampere circuit could be protected with a 20-ampere breaker if supplying only the fireplace, or with a 30-ampere breaker if other loads are connected to the same circuit. As with all equipment of this sort, grounding is essential.

## HUMIDIFIER

Portable humidifiers are designed to operate on 120 volts and are supplied with cord and grounding attachment plug so that they may be plugged into any convenient receptacle. Power consumption is low, from about 250 to 600 watts. They usually constitute a continuous load, so when determining the total load, multiply the nameplate rating by 1.25 and convert to amperage draw. Thus, a 500 watt unit on 120 volts would have a total demand of $500 \times 1.25 = 625$ watts, which divided by 120 equals 5.2 amperes. A humidifier can be used on any circuit, provided that the current rating of the circuit is not exceeded.

## EVAPORATIVE COOLER

Evaporative coolers are designed for operation at 120 volts, with a power consumption of 400 to 1000 watts. The same considerations apply as for humidifiers, except that larger models should be connected to individual or lightly loaded branch circuits.

## MICROWAVE OVEN

Microwave ovens have been perfected to the point where they are now reliable and coming into widespread use. They are supplied with cord and attachment plugs, operate on 120 volts, and consume from 1 kilowatts to 1.5 killowatts. They must be used with grounding receptacles and are best served by an individual 20-ampere branch circuit, or one that has only additional light loads or loads that will probably not operate at the same time as the oven. The attachment plug and receptacle arrangement serves as the disconnecting means.

# Glossary

accessible—1. As applied to wiring methods, capable of being removed or exposed without damaging the building structure or finish, or not permanently closed in by the structure or finish of the building. 2. As applied to equipment, admitting close approach because not guarded by locked doors, elevation or other effective means.

aggregate—Inert material mixed with cement and water to produce concrete.

appliance—Utilization equipment, generally equipment other than industrial, normally built in standardized sizes or types and installed or connected as a unit to perform one or more functions, such as clothes washing, air conditioning, food mixing, deep frying, etc.

appliance, fixed—An appliance that is fastened or otherwise secured at a specific location.

appliance, portable—An appliance that is actually moved or can easily be moved from one place to another in normal use.

appliance, stationary—An appliance that is not easily moved from one place to another in normal use.

approved—Acceptable to the authority enforcing the National Electrical Code.

attachment plug—A device that, upon insertion in a receptacle, established a connection between the conductors of the attached flexible cord and the conductors connected permanently to the receptacle.

**automatic**—Self-acting, operating by its own mechanism when actuated by some impersonal influence, such as a change in current strength, pressure, temperature, or mechanical configuration.

**backfill**—Loose earth placed outside foundation walls for filling and grading.

**bearing plate**—Steel plate placed under one end of a beam or truss for load distribution.

**bearing wall**—Wall supporting a load other than its own weight.

**bench mark**—Point of reference from which measurements are made.

**bonding jumper**—A reliable conductor used to insure the required electrical conductivity between metal parts required to be electrically connected.

**branch circuit**—That portion of a wiring system extending beyond the final overcurrent device protecting the circuit.

**branch circuit, appliance**—A circuit supplying energy to one or more outlets to which appliances are to be connected; such circuits have no permanently connected lighting fixtures that are not a part of an appliance.

**branch circuit, general-purpose**—A branch circuit that supplies a number of outlets for lighting and appliances.

**branch circuit, individual**—A branch circuit that supplies only one piece of utilization equipment.

**bridging**—System of bracing between floor beams to distribute floor load.

**building**—A structure that stands alone or that is cut off from adjoining structures by fire walls with all openings therein protected by approved fire doors.

**cabinet**—An enclosure designed for either surface or flush mounting and provided with a frame, mat, or trim in which swinging doors are hung.

**cavity wall**—Wall built of solid masonry units arranged to provide air space within the wall.

**chase**—Recess in inner face of masonry wall providing space for pipes or ducts.

**circuit breaker**—A device designed to open and close a circuit by nonautomatic means and to open the circuit automatically on a predetermined overload of current without injury to itself when properly applied within its rating.

**column**—Vertical load-carrying member of a structural frame.

**concealed**—Rendered inaccessible by the structure or finish of the building. Wires in concealed raceways are considered concealed, even though they may become accessible by withdrawing them.

**conductor, bare**—A conductor having no covering or insulation whatsoever.

**conductor, covered**—A conductor having one or more layers of nonconducting materials that are not recognized as insulation under the National Electrical Code.

**conductor, insulated**—A conductor covered with material recognized as insulation.

**connector, pressure**—A connector that establishes the connection between two or more conductors or between one or more conductors and a terminal by means of mechanical pressure and without the use of solder.

**continuous load**—A load in which the maximum current is expected to continue for 3 hours or more.

**contour line**—On a land map denoting elevations, a line connecting points with the same elevation.

**controller**—A device, or group of devices, that serves to govern in some predetermined manner, the electric power delivered to the apparatus to which it is connected.

**cooking unit, counter-mounted**—An assembly of one or more domestic surface heating elements for cooking purposes, designed to be flush mounted in, or supported by, a counter and complete with internal wiring and inherent or separately mounted controls.

**crawl space**—Shallow space between the first tier of beams and the ground (no basement).

**curtain wall**—Nonbearing wall between piers or columns for the enclosure of the structure; not supported at each story.

**demand factor**—In any system or part of a system, the ratio of the maximum demand of the system, or part of the system, to the total connected load of the system, or part of the system under consideration.

**disconnecting means**—A device, a group of devices, or other means whereby the conductors of a circuit can be disconnected from their source of supply.

**dry wall**—Interior wall construction consisting of plaster boards, wood paneling, or plywood nailed directly to the studs without application of plaster.

**duty, continuous**—A requirement of service that demands operation at a substantially constant load for an indefinitely long time.

**duty, intermittent**—A requirement of service that demands operation for alternate intervals of (1) load and no load, (2) load and rest or (3) load, no load and rest.

**duty, periodic**—A type of intermittent duty in which the load conditions regularly recur.

**duty, short-time**—A requirement of service that demands operations at loads and for intervals of time, both of which may be subject to wide variation.

**elevation**—Drawing showing the projection of a building on a vertical plane.

**enclosed**—Surrounded by a case that will prevent anyone from accidentally contacting live parts.

**equipment**—A general term including material, fittings, devices, appliances, fixtures, apparatus, and the like used as a part of, or in connection with, an electrical installation.

**expansion bolt**—Bolt with a casing arranged to wedge the bolt into a masonry wall to provide an anchorage.

**expansion joint**—Joint between two adjoining concrete members arranged to permit expansion and contraction with changes in temperature.

**exposed**—1. As applied to live parts, that which a person could inadvertently touch or approach nearer than a safe distance. This term is applied to parts not suitably guarded, isolated or insulated. 2. As applied to wiring method, not concealed.

**externally operable**—Capable of being operated without exposing the operator to contact with live parts.

**facade**—Main front of a building.

**feeder**—The conductors between the service equipment, or the generator switchboard of an isolated plant, and the branch-circuit overcurrent device.

**fire stop**—Incombustible filler material used to block interior draft spaces.

**fitting**—An accessory such as a locknut, bushing, or other part of a wiring system that is intended primarily to perform a mechanical rather than an electrical function.

**flashing**—Strips of sheet metal bent into an angle between the roof and wall to make a watertight joint.

**footing**—Structural unit used to distribute loads to the bearing materials.

**frost line**—Deepest level below grade to which frost penetrates in a geographic area.

**garage**—A building or portion of a building in which one or more self-propelled vehicles carrying volatile, flammable liquid for fuel or power are kept; also, all that portion of a building which is on or below the floor or floors in which such vehicles are kept and which is not separated from them by suitable cutoffs.

**ground**—A conducting connection, whether intentional or accidental, between an electrical circuit or a piece of equipment and earth or some other conducting body serving in place of the earth.

**grounded conductor**—A conductor used to connect equipment or the grounded circuit of a wiring system to a grounding electrode or electrodes.

**grounding conductor, main**—In an ungrounded system, the conductor connecting the equipment grounding conductor at the service to the grounding electrode.

**grounding conductor, equipment**—A conductor used to connect the equipment being grounded to the service-equipment enclosure.

**guarded**—Covered, shielded, fenced, enclosed, or otherwise protected by suitable covers or casings, barriers, rails or screens, mats or platforms in order to prevent dangerous contact or approach by persons or objects.

**I-beam**—Rolled steel beam or built-up beam of I-shaped configuration.

**Identified**—Used in the Code to refer to a conductor or terminal recognized as grounded.

**incombustible material**—Material that will not ignite or actively support combustion in a surrounding temperature of 1200 degrees Fahrenheit during an exposure of 5 minutes; also material that will not melt when the temperature of the material is maintained at 900 degrees Fahrenheit for a period of at least 5 minutes.

**isolated**—Not readily accessible to persons unless special means for access are used.

**jamb**—Upright member forming the side of a door or window opening.

**lally column**—Compression member consisting of a steel pipe filled with concrete under pressure.

**laminated wood**—Wood built up of plies or laminations that have been joined either with glue or with mechanical fasteners. Usually, the plies are too thick to be classified as veneer, and the grain of all plies is parallel.

**lighting outlet**—An outlet intended for the direct connection of a lamp holder, lighting fixture, or pendant cord terminating in a lamp holder.

**location, damp**—A location subject to a moderate amount of moisture, such as some basements, some barns, some cold-storage warehouses, etc.

**location, dry**—A location not normally subject to dampness or wetness. A location classified as dry may be temporarily subject to dampness or wetness, as in the case of a building under construction.

**location, wet**—A location subject to saturation with water or other liquids, such as locations exposed to weather, washrooms in garages, and similar locations. Installations that are located underground or in concrete slabs, or masonry in direct contact with the earth shall be considered wet locations.

**low-energy power circuit**—A circuit that is not a remote-control or signal circuit but whose power supply is limited in accordance with the requirements of class 2 remote-control circuits.

**multioutlet assembly**—A type of surface or flush raceway designed to hold conductors and attachment plug receptacles that is assembled in the field or at the factory.

**nonautomatic**—Used to describe an action requiring personal intervention for its control.

**nonbearing wall**—Wall that carries no load other than its own weight.

**outlet**—In the wiring system a point at which current is taken to supply utilization equipment.

**outline lighting**—An arrangement of incandescent lamps or gaseous tubes to outline and call attention to certain features such as the shape of a building or the decoration of a window.

**oven, wall-mounted**—A domestic oven for cooling purposes designed for mounting into or onto a wall or other surface.

**panelboard**—A single panel or group of panel units designed for assembly in the form of a single panel; includes buses and may come with or without switches and automatic overcurrent protective devices for the control of light, heat, or power circuits of small individual as well as aggregate capacity. It is designed to be placed in a cabinet or cutout box placed in or against a wall or partition and accessible only from the front.

**pilaster**—Flat square column attached to a wall and projecting about a fifth of its width from the face of the wall.

**plenum**—Chamber or space forming a part of an air-conditioning system.

**precast concrete**—Concrete units, such as piles or vaults, cast away from the construction site and set in place.

**qualified person**—One familiar with the construction and operation of the apparatus and the hazards involved.

**raceway**—Any channel designed expressly for holding wires, cables or bus bars and used solely for this purpose.

**rainproof**—So constructed, protected, or treated as to prevent rain from interfering with successful operation of the apparatus.

**raintight**—So constructed or protected that exposure to a beating rain will not result in the entrance of water.

**readily accessible**—Capable of being reached quickly for operation, renewal or inspections without requiring those to whom ready access is requisite to climb over or remove obstacles or resort to portable ladders, chairs, etc.

**receptacle outlet**—An outlet where one or more receptacles are installed.

**receptacle outlet, convenience**—A contact device installed at an outlet for the connection of an attachment plug.

**remote-control circuit**—Any electrical circuit that controls any other circuit through a relay or an equivalent device.

**riser**—Upright member of stair extending from tread to tread.

**roughing in**—Installation of all concealed electrical wiring; includes all electrical work done before finishing.

**sealed (hermetic-type) motor compressor**—A mechanical compressor consisting of a compressor and a motor, both of which are enclosed in the same sealed housing, with no external shaft or shaft seals, the motor operating in the refrigerant atmosphere.

**service**—The conductors and equipment used for delivering energy from the electricity supply system to the wiring system of the premises served.

**service cable**—The service conductors made up in the form of a cable.

**service conductors**—The supply conductors that extend from the street main or transformers to the service equipment of the premises being supplied.

**service drop**—The overhead service conductors from the last pole, or other aerial support, to and including the splices, if any, that connect to the service-entrance conductors at the building or other structure.

**service-entrance conductors, underground system**—The service conductors between the terminals of the service equipment and the point of connection to the service lateral.

**service equipment**—The necessary equipment, usually consisting of a circuit breaker, or switch and fuses and their accessories located near the point of entrance of supply conductors to a building and intended to constitute the main control and means of cutoff for the supply to that building.

**service lateral**—The underground service conductors between the street main, including any risers at a pole or other structure or from transformers, and the first point of connection to the service-entrance conductors in a terminal box, meter or other enclosure with adequate space, inside or outside the building wall. Where there is no terminal box, meter or other enclosure with adequate space the point of connection shall be considered to be the point of entrance of the service conductors into the building.

**service raceway**—The rigid metal conduit, electrical metallic tubing or other raceway that encloses the service-entrance conductors.

**setting, circuit breaker**—The value of the current at which the circuit breaker is set to trip.

**sheathing**—First covering of boards or paneling nailed to the outside of the wood studs of a frame building.

**siding**—Finishing material that is nailed to the sheathing of a wood frame building and that forms the exposed surface.

**signal circuit**—Any electrical circuit supplying energy to an appliance that gives a recognizable signal.

**soffit**—Underside of a stair, arch or cornice.

**soleplate**—Horizontal bottom member of wood stud partition.

**studs**—Vertically set skeleton members of a partition or wall to which lath is nailed.

**switch, general-use**—A switch intended for use in general distribution and branch circuits. It is rated in amperes and is capable of interrupting its rated voltage.

**switch, general-use snap**—A form of general-use switch so constructed that it can be installed in flush device boxes or on outlet covers, or otherwise used in conjunction with wiring systems recognized by this Code.

**switch, AC general-use snap**—A form of general-use snap switch suitable only for use on alternating-current circuits and for controlling the following: (1) resistive and inductive loads (including electric discharge lamps) not exceeding the ampere rating at the voltage involved, (2) tungsten-filament lamp loads not exceeding eighty percent of the ampere rating of the switches at the rated voltage.

**switch, AC-DC general-use snap**—A form of general use snap switch suitable for use on either direct- or alternating-current circuits and for controlling the following: (1) resistive loads not exceeding the ampere rating at the voltage involved, (2) inductive loads not exceeding one-half the ampere rating at the voltage involved, except that switches having a marked horsepower rating are suitable for controlling motors not exceeding the horsepower rating of the switch at the voltage involved and (3) tungsten-filament lamp loads not exceeding the ampere rating at 125 volts, when marked with the letter T.

**switch, isolating**—A switch intended for isolating an electric circuit from the source of power. It has no interrupting rating and is intended to be operated only after the circuit has been opened by some other means.

**switch, motor-circuit**—A switch, rated in horsepower, capable of interrupting the maximum operating overload current of a motor having the same horsepower rating as the switch at the rated voltage.

**switchboard**—A large single panel, frame or assembly of panels having switches and overcurrent protective devices, buses and usually instruments, mounted on the face or back or both. Switchboards are generally accessible from the rear as well as from the front and are not intended to be installed in cabinets.

**thermal cutout**—An overcurrent protective device containing a heater element in addition to and affecting a renewable fusible member which opens the circuit. It is not designed to interrupt short-circuit currents.

**thermally protected**—As applied to motors refers to the words thermally protected appearing on the nameplate of a motor or motor-compressor and means that the motor is provided with a thermal protector.

**thermal protector**—As applied to motors a protective device that is assembled as an integral part of a motor or motor-compressor and that, when properly applied, protects the motor against dangerous overheating due to overload and failure to start.

**trusses**—Framed structural pieces consisting of triangles in a single plane for supporting loads over spans.

**utilization equipment**—Equipment that utilizes electric energy for mechanical, chemical, heating, lighting or other similar useful purposes.

**ventilated**—Provided with a means to permit circulation of air sufficient to remove an excess of heat fumes or vapors.

**voltage**—The greatest root-mean-square (RMS), or effective, difference of potential between any two conductors of the circuit concerned.

**voltage to ground**—In grounded circuits the voltage between the given conductor and that point or conductor of the circuit which is grounded; in ungrounded circuits the greatest voltage between the given conductor and any other conductor of the circuit.

**watertight**—So constructed that moisture will not enter the enclosing case or housing.

**weatherproof**—So constructed or protected that exposure to the weather will not interfere with successful operation.

**web**—Central portion of an I-beam.

# Appendix I

# Manufacturers of

# Residential Electrical Products

This section lists the manufacturers of electrical equipment that is most often used in residential and farm wiring systems. Most of these manufacturers offer catalogs and other design data for their products—usually at no charge.

The homeowner and farmer will find the catalogs extremely helpful in selecting the proper equipment for a given project. Often the catalogs will contain data showing how to properly design an electrical system involving their products, and finally how to best install the system.

To use this section look under the heading material you wish to purchase. For example, suppose you would like information on the various outlet boxes available to install electrical wiring in an existing home. Look under the heading of *Boxes & Enclosures*, find the address of one or more manufacturers nearest you and write for data. After receiving the information you will be in a position to select the proper type for your situation by studying the description.

## ALARMS, SIGNALS & SYSTEMS

- Air King Corp.
  3050 N. Rockwell Ave.
  Chicago, IL 60618

- Artolier Lighting & Sound
  Div: Emerson Electric Co.
  141 Lanza Ave.
  Garfield, NJ 07026

- Auth Electric Co., Inc.
  Sub: Webster Electric Co.
  505 Acorn St.
  Deer Park, NY 11729

- Autocall
  Div: Federal Sign & Signal Corp
  Tucker Ave.
  Shelby, OH 44875

- Broan Mfg. Co.
  926 W. State St.
  Hardford, WI 53027

- Eagle Electric Mfg. Co., Inc.
  23–10 Bridge Plaza So.
  Long Island City, NY 11101

- Lutron Electronics Co., Inc.
  Sutter Rd.
  Coopersburg, PA 18036

- Nutone
  Div: Scovill Mfg. Co.
  Madison & Redbank Rds.
  Cincinnati, OH 45227

- Sierra Electric
  Div: Sola Basic Industries
  15100 S. Figueroa St.
  Gardena, CA 90247

- Thomas Industries, Inc.
  Div: Benjamin Products
  207 E. Broadway
  Louisville, KY 40202

## BOXES & ENCLOSURES

- Arrow-Hart, Inc.
  Murray Div.
  103 Hawthorn St.
  Hartford, CT 06106

- Bryant Electric
  Div: Westinghouse Electric
  1421 State St.
  Bridgeport, CT 06602

- Central Electric Products, Inc.
  1900 2nd Ave.
  Kearney, NE 68847

- Eagle Electric Mfg. Co., Inc.
  23–10 Bridge Plaza So.
  Long Island City, NY 11101

- Harvey Hubbell Inc.
  Div: Hubbell Pyle National
  1334 N. Kostner Ave.
  Chicago, IL 60651

- Pass & Seymour, Inc.
  50 Boyd Ave.
  Syracuse, NY 13209

- Raco
  Div: All-Steel Equipment, Inc.
  P.O. Box 871
  Aurora, IL 60507

- Stonco Lighting
  Div: Keene Corp.
  2345 Vauxhall Rd.
  Union, NJ 07083

- Wiremold Co.
  West Hartford, CT 06110

## CIRCUIT BREAKERS & FUSES

- Bussmann Mfg.
  Div: McGraw-Edison Co.
  University at Jefferson
  St. Louis, MO 63107

- Cutler-Hammer Inc.
  4201 N. 27th St.
  Milwaukee, WI 53216

- General Electric Co.
  Circuit Protective Devices
    Products Dept.
  41 Woodford Ave.
  Plainville, CT 06062

- Square D Co.
  Executive Plaza
  Park Ridge, IL 60068

■ Westinghouse
Div: Distribution Apparatus
P.O. Box 341
Bloomington, IN 47401

## COMMUNICATIONS

■ Air King Corp.
3050 N. Rockwell Ave.
Chicago, IL 60618

■ Altec
Div. LTV Ling Altec Inc.
1515 S. Manchester Ave.
Anaheim, CA 92803

■ Artolier Lighting & Sound
Div: Emerson Electric Co.
141 Lanza Ave.
Garfield, NJ 07026

Auth Electric Co. Inc.
Sub: Webster Electric Co.
505 Acorn St.
Deer Park, NY 11729

Bogen
Div: Lear Siegler, Inc.
P.O. Box 500
Paramus, NJ 07652

Couch, S.H.
Div: ESB Inc.
3 Arlington St.
N. Quincy, MA 02171

Executone Inc.
29–10 Thomson Ave.
Long Island City, NY 11101

Hemco Inc.
151–51 23 Ave.
Whitestone, NY 11357

Music & Sound Inc.
2961 Congressman Ln.
Dallas, TX 75220

Nutone
Div: Scovill Mfg. Co.
Madison & Redbank Rds.
Cincinnati, OH 45227

Talk-A-Phone Co.
5013 N. Kedzie Ave.
Chicago, IL 60625

Webster Electric Co. Inc.
Sub: Sta-Rite Industries
1900 Clark St.
Racine, WI 53403

## CONDUIT & RACEWAYS

■ General Electric Co.
Distribution Assemblies
Products Dept.
41 Woodford Ave.
Plainfille, CT 06062

■ Hatfield Wire & Cable
Div: Continental Copper &
Steel Ind. Inc.
360 Hurst St., Box 558
Linden, NJ 07036

Southwire Co.
Fertilla St.
Carrollton, GA 30117

Triangle Conduit & Cable Co., Inc.
Sub: Triangle Industries, Inc.
P.O. Box 711
New Brunswick, NJ 08903

Wheatland Tube Co.
Public Ledger Bldg.
Philadelphia, PA 19106

## CONNECTING WIRE &
## CABLE SPLICING DEVICES

■ Buchanan Electrical Products Corp.
Sub: Amerace Esna Corp.
1065 Floral Ave.
Union, NJ 07083

■ Burndy Corp.
Richards Ave.
Norwalk, CT 06856

■ 3M Co.
3-M Center
St. Paul, MN 55101

## FASTENERS, HANGERS, CLAMPS, TIES, ETC.

■ Ideal Industries, Inc.
5224 Becker Pl.
Sycamore, IL 60178

■ Rawplug Co., Inc., The
200 Petersville Rd.
New Rochelle, NY 10802

■ Thomas & Betts Co., The
Div: Thomas & Betts Corp.
36 Butler St.
Elizabeth, NJ 07207

## FITTINGS, CONDUIT & CABLE

■ Appleton Electric Co.
1701 Wellington Ave.
Chicago, IL 60657

■ Raco
Div: All-Steel Equipment, Inc.
P.O. Box 871
Aurora, IL 60507

Thomas & Betts Co., The
Div. Thomas & Betts Corp.
36 Butler St.
Elizabeth, NJ 07207

Triangle Conduit & Cable Co., Inc.
Sub: Triangle Industries, Inc.
P.O. Box 711
New Brunswick, NJ 08903

Wheatland Tube Co.
Public Ledger Bldg.
Philadelphia, PA 19106

## HEATING EQUIPMENT & CONTROLS

■ Chromalox Comfort Cond.
Div: Emerson Electric Co.
8100 Florissant
St. Louis, MO 63136

■ General Electric Co.
Industrial Heating Business Dept.
1 Progress Rd.
Shelbyville, IN 46176

Markel Electric Prods. Inc.
145 Seneca St.
Buffalo, NY 14203

The Singer Co.
Div: Climate Control
62 Columbus St.
Auburn, NY 13021

## LIGHT SOURCES

■ GTE Sylvania Inc.
100 Endicott St.
Danvers, MA 01923

■ General Electric Co.
Lamp Marketing Dept.
Nela Park
Cleveland, OH 44112

Westinghouse
Div: Lamp
1 Westinghouse Plaza
Bloomfield, NJ 07950

## LIGHTING FIXTURES, FLUORESCENT INDOORS

■ Markstone Mfg. Co.
Sub: Lightron Corp.
1240 N. Homan Ave.
Chicago, IL 60651

■ Progress Lighting
Div: LCA Corp.
Erie Ave. & G St.
Philadelphia, PA 19134

■ Swivelier Co., Inc.
33 Rt. 304
Nanuet, NY 10954

Thomas Industries, Inc.
Div: Residential Lighting
207 E. Broadway
Louisville, KY 40202

## SERVICE-ENTRANCE EQUIPMENT

■ Bryant Electric
Div: Westinghouse Electric
1421 State St.
Bridgeport, CT 06602

■ Cutler-Hammer Inc.
4201 N. 27th St.
Milwaukee, WI 53216

General Electric Co.
Circuit Protective Devices
  Prod. Dept.
41 Woodford Ave.
Plainville, CT 06062

I-T-E Imperial Corp.
233 E. Lancaster Ave.
Ardmore, PA 19003

■ Midwest Electric Mfg. Corp.
Sub: Crouse-Hinds Co.
1639 W. Walnut St.
Chicago, IL 60612

■ Square D Co.
Executive Plaza
Park Ridge, IL 60068

■ Westinghouse
Div: Distribution Control Equip.
Beaver, PA 15009

## SWITCHES, RECEPTACLES & PLATES

■ Arrow-Hart, Inc.
103 Hawthorn St.
Hartford, CT 06106

■ Bryant Electric
Div: Westinghouse Electric
1421 State St.
Bridgeport, CT 06602

Circle F Industries
720 Monmouth St.
Trenton, NJ 08604

General Electric Co.
General Purpose Control
  Prod. Dept.
P.O. Box 913
Bloomington, IL 61701

Harvey Hubbell Inc.
Div: Wiring Device
State St.
Bridgeport, CT 06602

Pass & Seymour, Inc.
50 Boyd Ave.
Syracuse, NY 13209

Sierra Electric
Div: Sola Basic Industries
15100 S. Figueroa St.
Gardena, CA 90248

Slater Electric Inc.
45 Sea Cliff Ave.
Glen Cove, NY 11542

Touch-Plate Electro Sys.
Sub: Circle F Industries
16530 Garfield Ave.
Paramount, CA 90723

Wiremold Co., The
West Hartford, CT 06110

## WIRE & CABLE

■ Hatfield Wire & Cable
Div: Continental Copper
  & Steel Inc. Inc.
360 Hurst St. Box 558
Linden, NJ 07036

■ Phelps Dodge Cable & Wire Co.
Div: Phelps Dodge Industries, Inc.
Foot of Point St.
Yonkers, NY 10702

Triangle Conduit & Cable Co., Inc.
Sub: Triangle Industries, Inc.
P.O. Box 711
New Brunswick, NJ 08903

## ELECTRICIANS' HAND TOOLS

■ Atkins Saw
Div: Nicholson File Co.
P.O. Box 958
Greenville, MS 38701

■ Benfield Benders
Appleton Electric Co.
Plastic Wire & Cable Corp.
1701 Wellington Ave.
Chicago, IL 60657

Holub Industries, Inc.
443 Elm St.
Sycamore, IL 60178

Ideal Industries, Inc.
5224 Becker Pl.
Sycamore, IL 60178

■ Klein & Sons Inc., Mathias
7200 McCormick Rd.
Chicago, IL 60645

## TEST & MEASURING DEVICES

■ Amprobe Instrument
Div. SOS Consolidated Inc.
630 Merrick Rd.
Lynbrook, NY 11563

■ Associated Research Inc.
3785 W. Belmont Ave.
Chicago, IL 60618

Biddle Co., James G.
Township Line & Jolly Roads
Plymouth Meeting, PA 19462

General Electric Co.
Utility & Process Automatic
 Products Dept.
40 Federal St.
Lynn MA 01910

Holub Industries, Inc.
443 Elm St.
Sycamore, IL 60178

Sperry Instrument Inc., A.W.
245 Marcus Pl.
Hauppauge, NY 11787

Westinghouse
Div: Electronic Tube
P.O. Box 284
Elmira, NY 14902

# Appendix II
# Wiring Tables

The wiring tables contained herein will aid you in laying out and installing many electrical circuits around the home and farm. For example, Table II-1 gives the recommended wire sizes for various loads anticipated on a circuit while other tables include those for calculating voltage drop.

Appendix II Table 1

| AWG/MCM | AMPERE RATING | | |
|---|---|---|---|
| | Type T, TW (60°C Wire) | Type RH, THWN, RHW, THW (75°C Wire) | Type RHH, THHN, XHHW (90°C Wire) |
| 14 | 15 | 15 | 15 |
| 12 | 20 | 20 | 20 |
| 10 | 30 | 30 | 30 |
| 8 | 40 | 45 | 50 |
| 6 | 55 | 65 | 70 |
| 4 | 70 | 85 | 90 |
| 3 | 80 | 100 | 105 |
| 2 | 95 | 115 | 120 |
| 1 | 110 | 130 | 140 |
| 0 | 125 | 150 | 155 |
| 00 | 145 | 175 | 185 |
| 000 | 165 | 200 | 210 |
| 0000 | 195 | 230 | 235 |
| 250M | 215 | 255 | 270 |
| 300M | 240 | 285 | 300 |
| 350M | 260 | 310 | 325 |
| 400M | 280 | 335 | 360 |
| 500M | 320 | 380 | 405 |
| 600M | 355 | 420 | 455 |
| 700M | 385 | 460 | 490 |
| 750M | 400 | 475 | 500 |
| 800M | 410 | 490 | 515 |
| 900M | 435 | 520 | 555 |
| 1000M | 455 | 545 | 585 |

## Appendix II Table 2 Electrical Symbols & Abbreviations

### GENERAL OUTLETS

CEILING WALL

| | | |
|---|---|---|
| O | –O | Outlet |
| Ⓑ | –Ⓑ | Blanked outlet |
| Ⓓ | | Drop cord |
| Ⓔ | –Ⓔ | Electrical outlet for use only when circle used alone might be confused with columns, plumbing symbols, etc. |
| Ⓕ | –Ⓕ | Fan outlet |
| Ⓙ | –Ⓙ | Junction box |
| Ⓛ | –Ⓛ | Junction box |
| Ⓛ$_{PS}$ | –Ⓛ$_{PS}$ | Lamp holder with pull switch |
| Ⓢ | –Ⓢ | Pull switch |
| Ⓥ | –Ⓥ | Outlet for vapor discharge lamp |
| Ⓧ | –Ⓧ | Exit light outlet |
| Ⓒ | –Ⓒ | Clock outlet (specify voltage) |

Lamp holder

### CONVENIENCE OUTLETS

| | |
|---|---|
| ⊜ | Duplex convenience outlet |
| ⊜$_{1,3}$ | Convenience outlet other than duplex 1 = single, 3 = triplex, etc. |
| ⊜$_{WP}$ | Weatherproof convenience outlet |
| ⊜$_R$ | Range outlet |
| ⊜$_S$ | Switch and convenience outlet |
| ⊜Ⓡ | Radio and convenience outlet |
| ▲ | Special purpose outlet. (des in spec) |
| ⊙ | Floor outlet |

### SWITCH OUTLETS

| | |
|---|---|
| S | Single pole switch |
| S$_2$ | Double pole switch |
| S$_3$ | Three-way switch |
| S$_4$ | Four-way switch |
| S$_D$ | Automatic door switch |
| S$_E$ | Electrolier switch |
| S$_K$ | Key operated switch |
| S$_P$ | Switch and pilot lamp |
| S$_{CB}$ | Circuit breaker |
| S$_{WCB}$ | Weatherproof circuit breaker |
| S$_{MC}$ | Momentary contact switch |
| S$_{RC}$ | Remote control switch |
| S$_{WP}$ | Weatherproof switch |
| S$_F$ | Fused switch |
| S$_{WF}$ | Weatherproof fused switch |

### SPECIAL OUTLETS

O$_{a,b,c,etc}$
⊜$_{a,b,c,etc}$
S$_{a,b,c,etc}$   Any standard symbol as given above with the addition of a lower case subscript letter may be used to designate some special variation of standard equipment of particular interest in a specific set of architectural plans. When used they must be listed in the key of symbols on each drawing and if necessary further described in the specifications.

## PANELS, CIRCUITS & MISCELLANEOUS

Lighting panel
Power panel
—— Branch circuit concealed in ceiling or wall
–·– Branch circuit concealed in floor
···· Branch circuit exposed

Home run to panelboard. Indicate number of circuits by number of arrows. Any circuit without further designation indicates a two-wire circuit. For a greater number of wires indicate as follows: (3 wires), (4 wires), etc.

—— Feeders. Use heavy lines and designate by number corresponding to listing in feeder schedule.

Underfloor duct and Junction box, triple system.
For double or single systems eliminate one or two lines.
This symbol is equally adaptable to auxiliary system layouts

(G) Generator
(M) Motor
(I) Instrument
(T) Power transformer (Or draw to scale)
⊠ Controller
Isolating switch

## AUXILIARY SYSTEMS

Pushbutton
Buzzer
Bell
Annunciator
Outside telephone
Interconnecting telephone
Telephone switchboard
(T) Bell ringing transformer
D Electric door opener
F Fire alarm bell
F Fire alarm station
City fire alarm station
FA Fire alarm central station
FS Automatic fire alarm device
W Watchman's station
W Watchman's central station
H Horn
N Nurse's signal plug
M Maid's signal plug
R Radio outlet
SC Signal central station
Interconnection box
Battery

—·—·— Auxiliary system circuits. Any line without further designation indicates a 2-wire system. For a greater number of wires designate with numerals in manner similar to—12-No. 18 W-¾" C., or designate by number corresponding to listing in schedule

□a,b,c Special auxiliary outlets. Subscript letters refer to notes on plans or detailed description in specifications.

| | 15 AMPERE | | 20 AMPERE | | 30 AMPERE | | 50 AMPERE | | 60 AMPERE | |
|---|---|---|---|---|---|---|---|---|---|---|
| | RECEPTACLE | PLUG | RECEPTACLE | PLUG | RECEPTACLE | PLUG | RECEPTACLE | PLUG | RECEPTACLE | PLUG |
| 1 — 125V | 1-15R | 1-15P | | | | | | | | |
| 2 — 250V | | 2-15P | 2-20R | 2-20P | 2-30R | 2-30P | | | | |
| 3 — 277V | | | | | | | | | | |
| 4 — 600V | | | | | | | | | | |
| 5 — 125V | 5-15R | 5-15P | 5-20R | 5-20P | 5-30R | 5-30P | 5-50R | 5-50P | | |
| 6 — 250V | 6-15R | 6-15P | 6-20R | 6-20P | 6-30R | 6-30P | 6-50R | 6-50P | | |
| 7 — 277V | 7-15R | 7-15P | 7-20R | 7-20P | 7-30R | 7-30P | 7-50R | 7-50P | | |
| 8 — 480V | | | | | | | | | | |
| 9 — 600V | | | | | | | | | | |
| 10 — 125/250V | | | 10-20R | 10-20P | 10-30R | 10-30P | 10-50R | 10-50P | | |

2-POLE 2-WIRE (rows 1–4)

2-POLE 3-WIRE GROUNDING (rows 5–10)

NEMA Wiring Device Configuration Chart

| | | 15R | 15P | 20R | 20P | 30R | 30P | 50R | 50P | 60R | 60P |
|---|---|---|---|---|---|---|---|---|---|---|---|
| **3-POLE 3-WIRE** | 11 — 3ϕ 250V | 11-15R | 11-15P | 11-20R | 11-20P | 11-30R | 11-30P | 11-50R | 11-50P | | |
| | — 3ϕ 480V | | | | | | | | | | |
| | 13 — 3ϕ 600V | | | | | | | | | | |
| **3-POLE 4-WIRE GROUNDING** | 14 — 125/250V | 14-15R | 14-15P | 14-20R | 14-20P | 14-30R | 14-30P | 14-50R | 14-50P | 14-60R | 14-60P |
| | 15 — 3ϕ 250V | 15-15R | 15-15P | 15-20R | 15-20P | 15-30R | 15-30P | 15-50R | 15-50P | 15-60R | 15-60P |
| | 16 — 3ϕ 480V | | | | | | | | | | |
| | 17 — 3ϕ 600V | | | | | | | | | | |
| **4-POLE 4-WIRE** | 18 — 3ϕ 120/208V | 18-15R | 18-15P | 18-20R | 18-20P | 18-30R | 18-30P | 18-50R | 18-50P | 18-60R | 18-60P |
| | 19 — 3ϕ 277/480V | | | | | | | | | | |
| | 20 — 3ϕ 347/600V | | | | | | | | | | |

341

# Appendix III
# Electrical Formulas

Several electrical formulas are contained herein to aid you in solving problems that might be encountered. Since inexpensive electronic calculators are now commonplace around every home and farm these examples have been set up for use with such devices.

## OHM'S LAW

Strickly speaking Ohms' law only applies to DC (direct current) circuits; however, in applications around the home and farm it can be applied to AC (alternating current) circuits as well. Ohm's law states that the voltage across a DC circuit is equal to the current time the resistance. That is,

$$E = IR$$

where $E$ is units volts, $I$ is in units of amperes and $R$ is in units of ohms.

## Example 1

Find the resistance when the current and voltage are known. The voltage is 120 volts and the current is 2 amperes. Hence,

1. Key in 120.
2. Press the division key.
3. Key in 2.
4. Press equals key.
5. Read answer—60 ohms.

## Example 2

Find the current when the resistance and voltage are known. The voltage is 240 volts and the resistance is 25 ohms. Hence,

1. Key in 240.
2. Press the division key.
3. Key in 25.
4. Press the equals key.
5. Read answer—9.6 amperes.

## Example 3

Find the voltage when the resistance and current are known. The resistance is 10 ohms and the current is 1.2 amperes, Hence,

1. Key in 10.
2. Press the multiplication key.
3. Key in 1.2.
4. Press the equals key.
5. Read answer—12 volts.

## POWER FORMULA

The formula for determining the DC power of an electrical circuit is calculated by multiplying the voltage across the circuit times the current. That is,

$$P = EI$$

where $P$ is in units of watts, $E$ is in units of volts and $I$ is in units of amperes.

From Ohm's law we know that the voltage is equal to the current times the resistance, and that current or resistance is

344

equal to the voltage divided by the other. It follows then that the voltage or current term in the above equation can be substituted with an equivalent expression, such as

$$P = E(E/R)$$
$$= E^2/R$$

or

$$P = (IR)I$$
$$= I^2R$$

In determining the power consumed by an AC circuit where the load is purely resistive, such as lamps and heating elements, this formula holds true; however, for applications where the load is inductive or capacitive a power factor must be introduced. Usually this number is stamped on the name plate of the device, such as a motor. Simply multiply the above equations by this factor. It is never greater than one and usually less.

## Example 1

Find the current consumed of a lamp when the power and voltage are known. The power is 100 watts and the voltage is 120 volts. Hence,

1. Key in 100.
2. Press the division key.
3. Key in 120.
4. Press the equals key.
5. Read answer—0.833...amperes.

## Example 2

Find the power consumed when the voltage and current are known. The voltage is 24 volts and the current 1.5 amperes. Hence,

1. Key in 24.
2. Press the multiplication key.
3. Key in 1.5.

4. Press the equals key.
5. Read answer—36 watts.

## Example 3

Find the voltage required when the power and current are known. The power is 500 watts and the current is 2 amperes. Hence,

1. Key in 500.
2. Press the division key.
3. Key in 2.
4. Press the equals key.
5. Read answer—250 volts.

## Example 4

Find the power consumed when the voltage and resistance are known. The voltage is 120 volts and the resistance is 100 ohms. Hence,

1. Key in 120.
2. Press the $x^2$ key.
3. Press the division key.
4. Key in 100.
5. Press the equals key.
6. Read answer—144 watts.

## Example 5

Find the power when the current and resistance are known. The current is 2 amperes and the resistance is 150 ohms. Hence,

1. Key in 2.
2. Press the $x^2$ key.
3. Press the multiplication key.
4. Key in 150.
5. Press the equals key.
6. Read answer—600 watts.

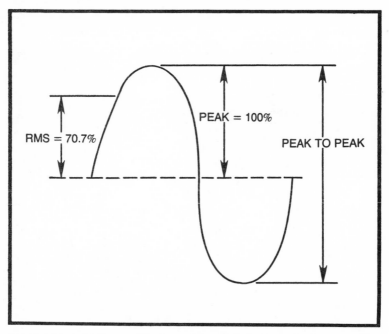

Fig. III-1. Typical AC wave. The type of wave is in the form of a sine wave. The RMS value is 0.707 of the peak value. The peak-to-peak value is twice that of the peak value.

## RMS VERSUS PEAK VALUE

AC voltage as its name implies is alternating current, alternating around zero or ground. Household current, 120 volts, is actually an effective value of the voltage present at each outlet. The peak value is around 169 volts. RMS, or root mean square, is 0.707 of the peak value. Refer to Fig. III-1 for a drawing of a typical AC wave.

## Example 1

Find the RMS value of an AC voltage peak. The peak value of the alternating current is 100 volts. Hence,

1. Key in 100.
2. Press the multiplication key.
3. Key in .707 (0.707).
4. Press the equals key.
5. Read answer—70.7 volts.

**Example 2**

Find the peak value of an AC voltage. The RMS value is 240 volts. Hence,

1. Key in 240.
2. Press the division key.
3. Key in .707 (0.707).
4. Press the equals key.
5. Read answer—339.46 volts.

# Appendix IV
# Lamp Data

The following listings are furnished by General Electric Company and contain data on electric lamps that are commonly used in the home and around the farm. Lamps in each group are listed in order of wattage, bulb size, voltage, description, approximate life, and the approximate initial lumens. The latter is necessary when using the lumens-per-square-foot method of calculating the recommended illumination levels in various areas as described in Chapter 12.

Besides those listed herein General Electric makes thousands of other types of lamps for specific lighting applications. For further information, consult your lamp supplier or write directly to General Electric, Nela Park, Cleveland, Ohio 44112.

Please note that these lamps will operate in any position except as noted in the tables. Also bear in mind that the prices listed will probably change by the time this goes to press, so verify them before ordering.

## INCANDESCENT LAMPS

| Bulb | Base | Lamp Ordering Code | Volts | Approx. List Price | DESCRIPTION | Std. Pkg. Qty. | Filament Design | M. O. L. | L. C. L. | Approx. Hours Life | Approx. Initial Lumens |
|---|---|---|---|---|---|---|---|---|---|---|---|
| **25 WATTS** | | | | | | | | | | | |
| F-10 | Cand. | **25F10C/F PM** | **115-125** | **$0.69** | Crystal (Clear) FLAIR | 120 | C-7A | 3 1/16 | 1 5/8 | 1500 | . . . |
| | | **25F10C/W/F PM** | **115-125** | **.79** | White FLAIR | 120 | C-7A | 3 1/16 | . . . | 1500 | . . . |
| F-15 | Medium | **25F/F PM** | **115-125** | **.57** | Crystal (Clear) FLAIR | 120 | C-9 | 4 3/8 | . . . | 1500 | 240 |
| | | **25F/W/F PM** | **115-125** | **.69** | White FLAIR | 120 | C-9 | 4 3/8 | . . . | 1500 | 179 |
| | | **25F/A/F PM** | **115-125** | **.69** | Colonial Amber FLAIR | 120 | C-9 | 4 3/8 | . . . | 1500 | . . . |
| | | **25F/AU/F PM** | **115-125** | **.69** | Auradescent FLAIR | 120 | C-9 | 4 3/8 | . . . | 1500 | . . . |
| G-16½ | Cand. | **25G16½C** | **115-125** | **.91** | Clear—Decorative | 60 | C-7A | 3 | 1 5/8 | 1500 | 235 |
| | | **25G16½C/W** | **115-125** | **1.16** | Sprayed White—Decorative | 60 | C-7A | 3 | . . . | 1500 | 210 |
| | | **25G16½C/W/F PM** | **115-125** | **1.16** | Sprayed White—Decorative FLAIR | 120 | C-7A | 3 | . . . | 1500 | 210 |
| G-18½ | Medium | **25G18½/W** | **115-125** | **.91** | Sprayed White—Decorative | 120 | C-9 | 3 9/16 | . . . | 1500 | 220 |
| G-25 | | **25G25** | **115-125** | **.81** | Clear—Decorative | 60 | C-9 | 4 7/16 | 2 9/16 | 1500 | 240 |
| | | **25G25/W** | **115-125** | **.91** | Sprayed White—Decorative | 60 | C-9 | 4 7/16 | . . . | 1500 | 220 |
| G-40 | ▲Medium | **25G40/CL** | **115-125** | **2.25** | Clear—Decorative (91) | 24 | C-9 | 6 13/16 | . . . | 2500 | 160 |
| | | **25G40/W** | **115-125** | **2.25** | White—Decorative (91) | 24 | C-9 | 6 13/16 | . . . | 2500 | 150 |
| A-19  | | **25A/B 24PK PM** | **115-125** | **.52** | Blue—Decorative. 24-Pack **PRICE MARKED** | 120 | C-9 | 3 7/8 | . . . | 2500 | . . . |
| | | **25A/G 24PK PM** | **115-125** | **.52** | Green—Decorative. 24-Pack **PRICE MARKED** | 120 | C-9 | 3 7/8 | . . . | 2500 | . . . |
| | | **25A/O** | **115-125** | **.52** | Orange--Decorative | 120 | C-9 | 3 7/8 | . . . | 2500 | . . . |
| | | **25A/R 24PK PM** | **115-125** | **.52** | Red—Decorative. 24-Pack **PRICE MARKED** | 120 | C-9 | 3 7/8 | . . . | 2500 | . . . |
| | | **25A/R2** | **115-125** | **.52** | Rose—Decorative | 120 | C-9 | 3 7/8 | . . . | 2500 | . . . |
| | | **25A/Y 24PK PM** | **115-125** | **.52** | Yellow—Decorative. 24-Pack **PRICE MARKED** | 120 | C-9 | 3 7/8 | . . . | 2500 | 180 |

| | | | | | | | | | | |
|---|---|---|---|---|---|---|---|---|---|---|
| | **25A19/GR/CL** | 130 | .48 | Clear—Sign. Group Replacem't | 120 | C-9 | 3⅞ | 2½ | 3000 | . . . |
| | **25A19/GR/IF** | 130 | .48 | Inside Frosted—Sign. Group Replacement | 120 | C-9 | 3⅞ | 2½ | 3000 | . . . |
| | **25A19/GR/TB** | 130 | .64 | Transparent Blue—Sign. Group Replacement (86) | 120 | C-9 | 3⅞ | 2½ | 3000 | . . . |
| | **25A19/TB 24PK PM** | 130 | .64 | Transparent Blue—Sign. Group Replacement. 24-Pack **PRICE MARKED (86)** | 120 | C-9 | 3⅞ | 2½ | 3000 | . . . |
| | **25A19/GR/TG** | 130 | .64 | Transparent Green—Sign. Group Replacement (86) | 120 | C-9 | 3⅞ | 2½ | 3000 | . . . |
| | **25A19/TG 24PK PM** | 130 | .64 | Transparent Green—Sign. Group Replacement. 24-Pack **PRICE MARKED** (86) | 120 | C-9 | 3⅞ | 2½ | 3000 | . . . |
| | **25A19/GR/TO** | 130 | .64 | Transparent Orange—Sign. Group Replacement (86) | 120 | C-9 | 3⅞ | 2½ | 3000 | . . . |
| | **25A19/TO 24PK PM** | 130 | .64 | Transparent Orange—Sign. Group Replacement. 24-Pack **PRICE MARKED** (86) | 120 | C-9 | 3⅞ | 2½ | 3000 | . . . |
| | **25A19/GR/TR** | 130 | .64 | Transparent Red—Sign. Group Replacement (86) | 120 | C-9 | 3⅞ | 2½ | 3000 | . . . |
| | **25A19/TR 24PK PM** | 130 | .64 | Transparent Red—Sign. Group Replacement. 24-Pack **PRICE MARKED** (86) | 120 | C-9 | 3⅞ | 2½ | 3000 | . . . |
| | **25A19/GR/TY** | 130 | .64 | Transparent Yellow—Sign. Group Replacement (86) | 120 | C-9 | 3⅞ | 2½ | 3000 | . . . |
| | **25A19/TY 24PK PM** | 130 | .64 | Transparent Yellow—Sign. Group Replacement. 24-Pack **PRICE MARKED** (86) | 120 | C-9 | 3⅞ | 2½ | 3000 | . . . |
| | **25A/NA** | 115-125 | 5.55 | Natural Amber | 120 | C-9 | 3⅞ | . . . | 2500 | . . . |
| | **25A/NB** | 115-125 | 5.55 | Natural Blue | 120 | C-9 | 3⅞ | . . . | 2500 | . . . |
| | **25A/NG** | 115-125 | 5.55 | Natural Green | 120 | C-9 | 3⅞ | . . . | 2500 | . . . |
| | **25A/NR** | 115-125 | 5.55 | Natural Ruby | 120 | C-9 | 3⅞ | . . . | 2500 | . . . |

All FLAIR Decorative Lamps are individually packaged and price marked, and are available in a 12-pack reshippable specialty pack.

# INCANDESCENT LAMPS

| Bulb | Base | Lamp Ordering Code | Volts | Approx. List Price | DESCRIPTION | Std. Pkg. Qty. | Filament Design | M. O. L. | L. C. L. | Approx. Hours Life | Approx. Initial Lumens |
|---|---|---|---|---|---|---|---|---|---|---|---|
| **25 WATTS (Continued)** | | | | | | | | | | | |
| A-21 | ▲Medium | **25A21/RFL** | **130** | **$1.87** | Reflector Sign—Light Inside Frosted | 120 | C-7A | 4⅜ | 1 31/32 | 2500 | . . . |
| G-16½ | S.C.Bay. | **25G16½SC** | **120** | **2.59** | Clear—Railway Light Signal (69) | 60 | C-5 | 3 | 1¼ | 1000 | . . . |
| S-6 | Cand. | **25S6** | **115-125** | **.60** | ✠ Clear—High Intensity Indicator. ANSI:DVN (16) | 240 | C-7A | 1⅞ | 1 5/16 | 50 | . . . |
| S-11 | | **25S11/2C** | **115-125** | **.88** | ✠ Clear—Medical Spotlight | 120 | C-7A | 2¼ | 1 7/16 | 500 | . . . |
| T-6½  | Intermed. | **25T6½** | **120** | **.99** | Clear—Showcase | 60 | C-8 | 5½ | . . . | 1000 | 244 |
| | | **25T6½ 12PK PM** | **120** | **.99** | Clear—Showcase. 12-Pack **PRICE MARKED** | 120 | C-8 | 5½ | . . . | 1000 | 244 |
| | | **25T6½** | **125-130** | **1.19** | Clear—Showcase | 60 | C-8 | 5½ | . . . | 1000 | 244 |
| | | **25T6½/IF** | **120** | **1.12** | Inside Frosted—Showcase | 60 | C-8 | 5½ | . . . | 1000 | 240 |
| | | **25T6½/IF** | **125-130** | **1.35** | Inside Frosted—Showcase | 60 | C-8 | 5½ | . . . | 1000 | 240 |
| | D.C.Bay. | **25T6½DC** | **120** | **1.42** | Clear—Appliance—Scale Illuminator | 60 | C-8 | 5 9/16 | . . . | 1000 | 244 |
| | | **25T6½DC** | **125-130** | **1.70** | Clear—Appliance—Scale Illuminator | 60 | C-8 | 5 9/16 | . . . | 1000 | 244 |
| | | **25T6½DC/IF** | **120** | **1.66** | Inside Frosted—Appliance—Scale Illuminator | 60 | C-8 | 5 9/16 | . . . | 1000 | 240 |
| | | **25T6½DC/IF** | **125-130** | **1.98** | Inside Frosted—Appliance—Scale Illuminator | 60 | C-8 | 5 9/16 | . . . | 1000 | 240 |
| T-8  | Cand. | **25T8C** | **115-125** | **.75** | Clear—Appliance | 60 | C-7A | 2⅝ | 1½ | (63) | . . . |
| | D.C.Bay. | **25T8DC** | **115-125** | **.57** | Clear—Home and Industrial Appliance | 60 | C-7A | 2⅝ | . . . | (63) | 240 |
| | | **25T8DC/PM** | **115-125** | **.69** | Clear—Appliance. Carded **PRICE MARKED** | 240 | C-7A | 2⅝ | . . . | (63) | 240 |

| | | | | | | | | | | | |
|---|---|---|---|---|---|---|---|---|---|---|---|
| | | **25T8DC/IF** | **115-125** | **.69** | Inside Frosted—Appliance | 60 | C-7A | 2⅝ | . . . | (63) | 235 |
| | Intermed. | **25T8N** | **115-125** | **.70** | Clear—Appliance | 60 | C-7A | 2⅝ | . . . | (63) | 240 |
| T-10  | Medium | **25T10** | **120** | **.64** | Clear—Showcase | 120 | C-8 | 5⅝ | . . . | 1000 | 248 |
| | | **25T10 24PK PM** | **120** | **.64** | Clear—Showcase. 24-Pack **PRICE MARKED** | 192 | C-8 | 5⅝ | . . . | 1000 | 248 |
| | | **25T10** | **125-130** | **.77** | Clear—Showcase | 120 | C-8 | 5⅝ | . . . | 1000 | 248 |
| | | **25T10/IF** | **120** | **.71** | Inside Frosted—Showcase | 120 | C-8 | 5⅝ | . . . | 1000 | 244 |
| | | **25T10/IF 24PK PM** | **120** | **.71** | Inside Frosted—Showcase. 24-Pack **PRICE MARKED** | 192 | C-8 | 5⅝ | . . . | 1000 | 244 |
| | | **25T10/IF** | **125-130** | **.85** | Inside Frosted—Showcase | 120 | C-8 | 5⅝ | . . . | 1000 | 244 |
| | Medium with Spring Contact | **25T10/RFL 24PK** | **120** | **2.97** | Reflector Showcase—Light Inside Frosted & side aluminized. M.O.L. is exclusive of spring contact on base. 24-Pack | 192 | CC-8 | 5⅝ | . . . | 1000 | 232 |
| | Medium | **25T10/6 PM** | **115-125** | **.92** | Clear—Aquarium **PRICE MARKED** | 24 | C-8 | 5⅝ | . . . | 2500 | . . . |
| R-14 | ▲Intermed. | **25R14N** | **120** | **1.89** | Reflector—Light Inside Frosted | 120 | CC-2V | 2 7/16 | . . . | 1500 | . . . |
| ★PAR-36 | Screw Terminal | **25PAR36** | **5.5** | **3.78** | PAR—Display Spotlight. Filament shielded | 12 | C-6 | 2¾ | . . . | 1000 | . . . |
| ★PAR-46 | | **25PAR46** | **5.5** | **3.78** | PAR—Display Spotlight. Filament shielded | 12 | C-6 | 3¾ | . . . | 1000 | . . . |
| S-11 | Long Shell Medium | **25S11/7** | **6** | **2.16** | Clear—Instrument. Source WxH: 4.4x0.8mm. Filament 5/16" max. from end of bulb (31,76) | 120 | C-6 | 2⅝ | . . . | 50 | . . . |
| T-8 | S.C.Pref. (A) | **25T8/SCP** | **6** | **2.48** | Clear—Instrument. Source WxH: 0.8x4.5mm. (57,61) | 24 | C-8 | 3⅛ | 1⅝ | 50 | . . . |
| S-11 | S.C.Bay. | **25S11/4SC** | **10** | **1.89** | Clear—Railway Light Signal (70) | 120 | CC-6 | 2⅜ | 1¼ | 1000 | . . . |
| A-19 | Medium | **25A** | **12** | **.71** | Inside Frosted (11) | 120 | C-6 | 3⅞ | 2 9/16 | 1000 | 378 |
| ★PAR-36 | Screw Terminal | **25PAR36NSP** | **12** | **3.45** | PAR—Narrow Spotlight (65) | 12 | C-6 | 2¾ | . . . | 2000 | . . . |
| | | **25PAR36WFL** | **12** | **3.45** | PAR—Wide Floodlight (65) | 12 | C-6 | 2¾ | . . . | 2000 | . . . |
| | | **25PAR36/VWFL** | **12** | **3.45** | PAR—Very Wide Floodlight (65) | 12 | C-6 | 2¾ | . . . | 2000 | . . . |

# INCANDESCENT LAMPS

| Bulb | Base | Lamp Ordering Code | Volts | Approx. List Price | DESCRIPTION | Std. Pkg. Qty. | Filament Design | M. O. L. | L. C. L. | Approx. Hours Life | Approx. Initial Lumens |
|---|---|---|---|---|---|---|---|---|---|---|---|
| **25 WATTS (Continued)** | | | | | | | | | | | |
| R-14  | S.C.Bay. | 25R14SC/SP | 12 | $1.99 | Reflector Spotlight—Light Inside Frosted (15) | 120 | CC-8 | $2\frac{5}{8}$ | ... | 2000 | ... |
| R-14 | S.C.Bay. | 25R14SC/FL | 12 | 1.99 | Reflector Floodlight—I.F. (15) | 120 | CC-8 | $2\frac{5}{8}$ | ... | 2000 | ... |
| R-14 | S.C.Bay. | 25R14SC | 28 | 2.75 | Reflector Spotlight—Light Inside Frosted (15) | 120 | CC-8 | $2\frac{5}{8}$ | ... | 2000 | ... |
| A-19 | Medium | 25A | 30 | .59 | Inside Frosted—Train (11) | 120 | C-9 | $3\frac{7}{8}$ | $2\frac{9}{16}$ | 1000 | 345 |
| T-8½ | Medium | 25T8½/IF | 34 | 1.94 | Inside Frosted—Train | 60 | C-8 | $5\frac{3}{8}$ | ... | 1000 | ... |
| T-8½ | Medium | 25T8½/IF | 60 | 2.32 | Inside Frosted—Train | 60 | C-8 | $5\frac{3}{8}$ | ... | 1000 | ... |
| A-19 | Medium | 25A | 34 | .64 | Inside Frosted—Train (11) | 120 | C-9 | $3\frac{7}{8}$ | $2\frac{9}{16}$ | 1000 | 390 |
| ★PAR-38 | Medium Skirted | 25PAR/FL | 34 | 3.45 | PAR—Train—Floodlight | 12 | C-6 | $5\frac{5}{16}$ | ... | 1000 | ... |
| ★PAR-38 | Medium Skirted | 25PAR/FL | 60 | 3.45 | PAR—Train—Floodlight | 12 | C-6 | $5\frac{5}{16}$ | ... | 1000 | 180 |
| A-19 | Medium | 25A | 60 | .47 | Inside Frosted—Train (11) | 120 | C-9 | $3\frac{7}{8}$ | $2\frac{9}{16}$ | 1000 | 285 |
| A-17 | Medium | 25A17/RS | 75 | .68 | Inside Frosted—Train. Rough Service | 120 | C-9 | $3\frac{5}{8}$ | $2\frac{1}{2}$ | 1000 | 250 |
| A-19 | Medium | 25A | 230,250 | .71 | Inside Frosted | 120 | C-17A | $3\frac{7}{8}$ | $2\frac{9}{16}$ | 1000 | 220 |
| A-19 | Medium | 25A | 300 | .83 | Inside Frosted | 120 | C-17A | $3\frac{7}{8}$ | $2\frac{9}{16}$ | 1000 | ... |
| A-19 | ▲Medium | 25A/R | 230,250 | .83 | Red | 120 | C-17A | $3\frac{7}{8}$ | ... | 1000 | ... |
| T-6½ | D.C.Bay. | 25T6½DC/IF | 230 | 1.72 | Inside Frosted—Scale Illuminator | 60 | C-8 | $5\frac{9}{16}$ | ... | 1000 | ... |
| T-10 | Medium | 25T10 24PK | 230,250 | 1.18 | Clear—Showcase. 24-Pack | 192 | C-17A | $5\frac{5}{8}$ | ... | 1000 | ... |
| **30 WATTS** | | | | | | | | | | | |
| R-20 | ▲Medium | 30R20 | 115-125 | 1.93 | Reflector—Light Inside Frosted | 60 | C-7A | $3\frac{15}{16}$ | ... | 2000 | 210 |
| R-20 | ▲Medium | 30R20 | 130 | 2.32 | Reflector—Light Inside Frosted | 60 | C-7A | $3\frac{15}{16}$ | ... | 2000 | 210 |

| | | | | | | | | | | |
|---|---|---|---|---|---|---|---|---|---|---|
| G-16½ | D.C.Bay. | **30G16½/3DC** | **115** | **5.00** | Clear—Medical Spotlight. Silvered Bowl | 60 | CC-2V Horiz. | 3 5/16 | 1 13/16 | 500 | . . . |
| T-8 | Disc | **L30** | **115-125** | **3.34** | Clear—Lumiline | 24 | C-8 | 17 3/4° | . . . | 1500 | . . . |
| | | **L30/IF** | **115-125** | **3.56** | Inside Frosted—Lumiline | 24 | C-8 | 17 3/4° | . . . | 1500 | 255 |
| | | **L30/W** | **115-125** | **3.56** | White—Lumiline | 24 | C-8 | 17 3/4° | . . . | 1500 | 210 |
| S-11 | S.C.Pref. (A) | **30S11/4SCP** | **6** | **2.70** | Clear—Photo-micrographic. Source WxH: 2.1x1.9mm. Filament 5/16" max. from end of bulb (31,76) | 120 | C-6 | 2 3/8 | 1 5/8 | 500 | . . . |
| | German Cand. | **30S11/10** | **6** | **2.59** | Clear—Instrument. Source WxH: 3.2x1.0mm. (31) | 120 | C-6 | 2 1/2 | 1 13/16 | 50 | . . . |
| | D.C.Bay. | **30S11/100** | **32** | **.97** | Clear—Train Marker (57) | 120 | C-7A | 2 3/8 | 1 1/4 | 500 | . . . |
| T-8 | Disc | **L30/W** | **60** | **3.44** | White—Lumiline | 24 | C-8 | 11 3/4° | . . . | 1500 | . . . |
| S-11 | D.C.Bay. | **30S11DC** | **64** | **.86** | Clear—Train Marker (57) | 120 | C-7A | 2 3/8 | 1 1/4 | 500 | 350 |
| | | **30S11DC** | **75** | **.86** | Clear—Train Marker Control Stand (57) | 120 | C-7A | 2 3/8 | 1 1/4 | 500 | . . . |

## 30-70-100 WATTS

| | | | | | | | | | | | |
|---|---|---|---|---|---|---|---|---|---|---|---|
| A-21 | 3-Contact Medium | **30/100/W** **12PK PM** | **120** | **.89** | Soft-White—3-Way. Burn base down. 12-Pack. **PRICE MARKED** | 60 | C-2R CC-8 | 5 1/4 | 3 3/4 | 1500, 1200, 1150 | 280, 1035, 1315 |

## 36 WATTS

| | | | | | | | | | | | |
|---|---|---|---|---|---|---|---|---|---|---|---|
| A-19 | Medium | **36A/RYH** | **120** | **1.29** | Street Railway Headlight. Design amps: .342 (41) | 120 | C-5 | 3 7/8 | 2 1/4 | 1000 | . . . |

## 40 WATTS

| | | | | | | | | | | | |
|---|---|---|---|---|---|---|---|---|---|---|---|
| A-15 | Medium | **40A15** | **115-125** | **.33** | Inside Frosted—Appliance. Oven (11) | 120 | C-9 | 3 1/2 | 2 3/8 | 1500 | 455 |
| | | **40A15/PM** | **115-125** | **.45** | Inside Frosted—Appliance. Oven. Carded. **PRICE MARKED** (11) | 240 | C-9 | 3 1/2 | 2 3/8 | 1500 | 455 |

# INCANDESCENT LAMPS

## 40 WATTS (Continued)

| Bulb | Base | Lamp Ordering Code | Volts | Approx. List Price | DESCRIPTION | Std. Pkg. Qty. | Filament Design | M. O. L. | L. C. L. | Approx. Hours Life | Approx. Initial Lumens |
|---|---|---|---|---|---|---|---|---|---|---|---|
| A-19  | Medium | **40A** | **120** | **$0.34** | Inside Frosted (11) | 120 | C-9 | 4¼ | 2 15/16 | 1500 | 455 |
| | | **40A 24PK PM** | **120** | **.34** | Inside Frosted. 24-Pack. **PRICE MARKED** | 120 | C-9 | 4¼ | 2 15/16 | 1500 | 455 |
| | | **40A/TF 24PK** | **115-125** | **.76** | Inside Frosted—TUFF-SKIN 24-Pack | 120 | C-9 | 4¼ | . . . | 1500 | . . . |
| | | **40A** | **130** | **.42** | Inside Frosted (11) | 120 | C-9 | 4¼ | 2 15/16 | 1500 | 455 |
| | | **40A/W 24PK PM** | **120** | **.37** | Soft-White. 24-Pack. **PRICE MARKED** (11) | 120 | C-9 | 4¼ | 2 15/16 | 1500 | 440 |
| | | **40A/CL** | **120** | **.39** | Clear (11) | 120 | C-9 | 4¼ | 2 15/16 | 1500 | 455 |
| | | **40A/CL** | **130** | **.47** | Clear (11) | 120 | C-9 | 4¼ | 2 15/16 | 1500 | 455 |
| | ▲Medium | **40A/99** | **120** | **.48** | Inside Frosted—Extended Service (11) | 120 | C-9 | 4¼ | 2 15/16 | 2500 | 420 |
| | | **40A/99 24PK** | **120** | **.48** | Inside Frosted—Extended Service. 24-Pack (11) | 120 | C-9 | 4¼ | 2 15/16 | 2500 | 420 |
| | | **40A/99** | **130** | **.57** | Inside Frosted—Extended Service (11) | 120 | C-9 | 4¼ | 2 15/16 | 2500 | 420 |
| | Medium | **40A/DPK 24PK PM** | **115-125** | **.52** | Coloramic—Dawn Pink. 24-Pack **PRICE MARKED** (11) | 120 | C-9 | 4¼ | . . . | 1500 | . . . |
| B-10  | Cand. | **40B10/F PM** | **115-125** | **.79** | Crystal (Clear). Burn base down to horizontal. FLAIR | 120 | C-9 | 3¾ | 1⅞ | 1500 | . . . |
| | | **40B10/W/F PM** | **115-125** | **.89** | Satin White. Burn base down to horizontal. FLAIR | 120 | C-9 | 3¾ | . . . | 1500 | . . . |
| B-13 | Medium | **40B13/F PM** | **115-125** | **.79** | Crystal (Clear) FLAIR | 120 | C-9 | 4⅝ | 2⅞ | 1500 | 425 |
| | | **40B13/W/F PM** | **115-125** | **.89** | Satin White FLAIR | 120 | C-9 | 4⅝ | . . . | 1500 | 385 |
| CA-9 | ▲Medium | **40CA9/F PM** | **115-125** | **.79** | Clear—Bent Tip. (85) FLAIR | 120 | CC-2V | 4½ | 2 9/16 | 1500 | . . . |
| | | **40CA9/2/F PM** | **115-125** | **.89** | Outside Frosted—Bent Tip (85) FLAIR | 120 | CC-2V | 4½ | 2 9/16 | 1500 | . . . |

| | | | | | | | | | | | | |
|---|---|---|---|---|---|---|---|---|---|---|---|---|
| CA-10 | ▲Cand. | 40CA10/F | PM | 115-125 | .79 | Clear—Bent Tip (85) FLAIR | 120 | CC-2V | 4 5/16 | 2 1/8 | 1500 | . . |
| | | 40CA10/2/F | PM | 115-125 | .89 | Outside Frosted—Bent Tip (85) FLAIR | 120 | CC-2V | 4 5/16 | 2 1/8 | 1500 | . . . |
| F-15 | Medium | 40F15/F | PM | 115-125 | .69 | Crystal (Clear) FLAIR | 120 | C-9 | 4 3/8 | . . . | 1500 | 435 |
| | | 40F15/W/F | PM | 115-125 | .79 | White FLAIR | 120 | C-9 | 4 3/8 | . . . | 1500 | 390 |
| | | 40F15/AU/F | PM | 115-125 | .79 | Auradescent FLAIR | 120 | C-9 | 4 3/8 | . . . | 1500 | . . . |
| G-16½ | Cand. | 40G16½C/W | | 115-125 | 1.60 | Sprayed White—Decorative. Burn base down to horizontal | 60 | C-7A | 3 | . . . | 1500 | 245 |
| | D.C.Bay. | 40G16½/3DC | | 115 | 6.40 | Clear—Silvered Bowl. Medical Spotlight | 60 | CC-2V Horiz. | 3 5/16 | 1 13/16 | 500 | . . . |
| G25 | Medium | 40G/W | | 115-125 | .92 | Sprayed White—Decorative | 60 | C-9 | 4 7/16 | . . . | 1500 | 310 |
| G-40 | ▲Medium | 40G40/CL | | 115-125 | 1.82 | Clear—Decorative | 24 | C-9 | 6 13/16 | . . . | 2500 | 385 |
| | | 40G40/W/F | PM | 115-125 | 2.25 | White—Moonglow FLAIR | 24 | C-9 | 6 13/16 | . . . | 2500 | 355 |
| | | 40G40/W | | 130 | 2.25 | White—Decorative | 24 | C-9 | 6 13/16 | . . . | 2500 | 355 |
| A-21  | | 40A/B | | 115-125 | .62 | Blue—Decorative | 120 | C-9 | 4 3/8 | . . . | 1000 | . . . |
| | | 40A/G | | 115-125 | .62 | Green—Decorative | 120 | C-9 | 4 3/8 | . . . | 1000 | . . . |
| | | 40A/O | | 115-125 | .62 | Orange—Decorative | 120 | C-9 | 4 3/8 | . . . | 1000 | . . . |
| | | 40A/R | | 115-125 | .62 | Red—Decorative | 120 | C-9 | 4 3/8 | . . . | 1000 | . . . |
| | | 40A/Y | | 115-125 | .62 | Yellow—Decorative | 120 | C-9 | 4 3/8 | . . . | 1000 | 290 |
| | | 40A/Y | | 130 | .74 | Yellow—Decorative | 120 | C-9 | 4 3/8 | . . . | 1000 | 290 |
| | | 40A21/GR/CL | | 130 | .48 | Clear—Sign. Group Replacem't | 120 | C-9 | 4 3/8 | 2 15/16 | 3000 | . . . |
| | | 40A21/GR/IF | | 130 | .48 | Inside Frosted—Sign. Group Replacement | 120 | C-9 | 4 3/8 | 2 15/16 | 3000 | . . . |
| | | 40A21/GR/TB | | 130 | .70 | Transparent Blue—Sign. Group Replacement (86) | 120 | C-9 | 4 3/8 | 2 15/16 | 3000 | . . . |
| | | 40A21/GR/TG | | 130 | .70 | Transparent Green—Sign. Group Replacement (86) | 120 | C-9 | 4 3/8 | 2 15/16 | 3000 | . . . |
| | | 40A21/GR/TO | | 130 | .70 | Transparent Orange—Sign. Group Replacement (86) | 120 | C-9 | 4 3/8 | 2 15/16 | 3000 | . . . |

All FLAIR Decorative Lamps are individually packaged, price marked and are available in a 12-pack reshippable specialty pack.

## INCANDESCENT LAMPS

| Bulb | Base | Lamp Ordering Code | Volts | Approx. List Price | DESCRIPTION | Std. Pkg. Qty. | Filament Design | M. O. L. | L. C. L. | Approx. Hours Life | Approx. Initial Lumens |
|---|---|---|---|---|---|---|---|---|---|---|---|
| **40 WATTS** (Continued) | | | | | | | | | | | |
| A-21  | ▲Medium | **40A21/GR/TR** | **130** | **$0.70** | Transparent Red—Sign. Group Replacement (86) | 120 | C-9 | 4⅜ | 2 15/16 | 3000 | . . . |
| | | **40A21/GR/TY** | **130** | **.70** | Transparent Yellow—Sign. Group Replacement (86) | 120 | C-9 | 4⅜ | 2 15/16 | 3000 | . . . |
| | | **40A/TS** | **120** | **.47** | Clear—Traffic Signal (57) | 120 | C-9 | 4⅜ | 2 7/16 | 2000 | 380 |
| | | **40A/TS** | **130** | **.56** | Clear—Traffic Signal (57) | 120 | C-9 | 4⅜ | 2 7/16 | 2000 | 380 |
| T-6½ | Intermed. | **40T6½/2** | **115-125** | **1.18** | Clear—Refrigerator | 60 | C-8 | 5½ | . . . | 750 | . . . |
| | | **40T6½/2IF** | **115-125** | **1.29** | Inside Frosted—Appliance | 60 | C-8 | 5½ | . . . | 750 | . . . |
| T-8 | Medium | **40T8** | **120** | **2.09** | Clear—Showcase | 24 | C-23 | 11⅞ | . . . | 1000 | 430 |
| | | **40T8** | **125-130** | **2.49** | Clear—Showcase | 24 | C-23 | 11⅞ | . . . | 1000 | 430 |
| | | **40T8/IF** | **120** | **2.37** | Inside Frosted—Showcase | 24 | C-23 | 11⅞ | . . . | 1000 | 425 |
| | | **40T8/IF** | **125-130** | **2.85** | Inside Frosted—Showcase | 24 | C-23 | 11⅞ | . . . | 1000 | 425 |
| T-10  | | **40T10** | **120** | **.64** | Clear—Showcase | 120 | C-8 | 5⅝ | . . . | 1000 | 450 |
| | | **40T10 24PK PM** | **120** | **.64** | Clear—Showcase. 24-Pack **PRICE MARKED** | 192 | C-8 | 5⅝ | . . . | 1000 | 450 |
| | | **40T10** | **125-130** | **.77** | Clear—Showcase | 120 | C-8 | 5⅝ | . . . | 1000 | 450 |
| | | **40T10/IF** | **120** | **.69** | Inside Frosted—Showcase | 120 | C-8 | 5⅝ | . . . | 1000 | 445 |
| | | **40T10/IF 24PK PM** | **120** | **.69** | Inside Frosted—Showcase 24-Pack **PRICE MARKED** | 192 | C-8 | 5⅝ | . . . | 1000 | 445 |
| | | **40T10/IF** | **125-130** | **.83** | Inside Frosted—Showcase | 120 | C-8 | 5⅝ | . . . | 1000 | 445 |
| | Medium with Spring Contact | **40T10/RFL 24PK** | **120** | **3.34** | Reflector Showcase—Light Inside Frosted & side aluminized. M.O.L. is exclusive of spring contact on base. 24-Pack | 192 | CC-8 | 5⅝ | . . . | 1000 | 430 |

| Shape | Base | Ordering Code | Volts | | Description | | Filament | M.O.L. | L.C.L. | Rated Life | Lumens |
|---|---|---|---|---|---|---|---|---|---|---|---|
| T-8 | Medium Prefocus | 40T10P | 120 | 2.91 | Clear—Airport (1) | 60 | CC-2V | 3 9/16 | 1 1/2 | 1000 | ... |
| T-8 | Disc | L40 | 115-125 | 2.90 | Clear—Lumiline | 24 | C-8 | 11 3/4° | ... | 1500 | ... |
| | | L40/IF | 115-125 | 3.01 | Inside Frosted—Lumiline | 24 | C-8 | 11 3/4° | ... | 1500 | 340 |
| | | L40/W | 115-125 | 3.01 | White—Lumiline | 24 | C-8 | 11 3/4° | ... | 1500 | 295 |
| A-15 | 2-Lug Sleeve | 40A15/2 | 10 | 3.24 | Clear—Railway Light Signal (10,11) | 120 | C-2V | 3 3/4 | 2 1/32 | 1500 | ... |
| A-15 | 3-Lug Sleeve | 40A15/3 | 10 | 3.32 | Clear—Railway Light Signal (10,11) | 120 | C-2V | 3 3/4 | 2 3/8 | 1500 | ... |
| T-8 1/2 | Medium | 40T8 1/2/IF | 30 | 1.83 | Inside Frosted—Train | 60 | C-8 | 5 3/4 | ... | 1000 | ... |
| A-19 | | 40A | 60 | .70 | Inside Frosted—Train (11) | 120 | C-9 | 4 1/4 | 2 9/16 | 1000 | ... |
| T-10 | | 40T10 | 24PK 230 | 1.29 | Clear—Showcase. 24-Pack | 192 | C-17A | 5 3/4 | ... | 1000 | ... |

## 45 WATTS

| Shape | Base | Ordering Code | Volts | | Description | | Filament | M.O.L. | L.C.L. | Rated Life | Lumens |
|---|---|---|---|---|---|---|---|---|---|---|---|
| ★PAR-38 | Medium Skirted | 45PAR38/6.6 | 6.6A | 4.86 | PAR—Airport—Floodlight | 12 | C-6 | 5 7/16 | ... | 800 | ... |

## 50 WATTS

| Shape | Base | Ordering Code | Volts | | Description | | Filament | M.O.L. | L.C.L. | Rated Life | Lumens |
|---|---|---|---|---|---|---|---|---|---|---|---|
| A-19 | Medium | 50A/RS | 120 | .71 | Inside Frosted—Rough Service | 120 | C-22 | 3 7/8 | 2 9/16 | 1000 | 480 |
| | | 50A/RS | 24PK 120 | .71 | Inside Frosted—Rough Service. 24-Pack | 120 | C-22 | 3 7/8 | 2 9/16 | 1000 | 480 |
| | | 50A/RS/TF | 24PK 115-125 | 1.12 | Inside Frosted—Rough Service. TUFF-SKIN 24-Pack | 120 | C-22 | 3 7/8 | ... | 1000 | ... |
| | | 50A/RS | 130 | .85 | Inside Frosted—Rough Service | 120 | C-22 | 3 7/8 | 2 9/16 | 1000 | 480 |
| | | 50A/RS | 24PK 130 | .85 | Inside Frosted—Rough Service 24-Pack | 120 | C-22 | 3 7/8 | 2 9/16 | 1000 | 480 |
| | | 50A19/5 | 120 | .75 | Clear—Rough Service | 120 | C-22 | 3 7/8 | 2 9/16 | 1000 | ... |
| | | 50A19/5 | 130 | .90 | Clear—Rough Service | 120 | C-22 | 3 7/8 | 2 9/16 | 1000 | ... |
| | | 50A/VS | 24PK 120 | .50 | Inside Frosted—Vibration Service 24-Pack | 120 | C-9 | 3 7/8 | 2 1/2 | 1000 | 550 |
| | | 50A/VS | 24PK 130 | .61 | Inside Frosted—Vibration Service 24-Pack | 120 | C-9 | 3 7/8 | 2 1/2 | 1000 | 550 |

## INCANDESCENT LAMPS

| Bulb | Base | Lamp Ordering Code | Volts | Approx. List Price | DESCRIPTION | Std. Pkg. Qty. | Filament Design | M. O. L. | L. C. L. | Approx. Hours Life | Approx. Initial Lumens |
|---|---|---|---|---|---|---|---|---|---|---|---|
| **50 WATTS (Continued)** | | | | | | | | | | | |
| GA-25  | Medium | **50GA** | 115-125 | $0.91 | Inside Frosted—Ivory bowl. Semi-indirect (3) | 24 | C-9 | 4 7/16 | . . . | 1000 | 575 |
| | | **50GA/DPK** | 115-125 | .91 | Coloramic—Inside Frosted. Dawn Pink Bowl. Semi-indirect (3) | 24 | C-9 | 4 7/16 | . . . | 1000 | . . . |
| R-20  | ▲Medium | **50R20** | 115-125 | 2.10 | Reflector—Light Inside Frosted (14) | 60 | C-7A | 3 15/16 | . . . | 2000 | 440 |
| | | **50R20** | 130 | 2.53 | Reflector—Light Inside Frosted (14) | 60 | C-7A | 3 15/16 | . . . | 2000 | 440 |
| | | **50R20BW** | 115-125 | 2.90 | Reflector—Blue White (14 34, 35) | 60 | C-7A | 3 15/16 | . . . | 2000 | . . . |
| | | **50R20PK** | 115-125 | 2.90 | Reflector—Pink (14,34,35) | 60 | C-7A | 3 15/16 | . . . | 2000 | . . . |
| R-30 | | **50R30/FL** | 120 | 1.62 | Reflector Floodlight—Inside Frosted (14) | 24 | CC-6 | 5 3/8 | . . . | 2000 | . . . |
| A-19 | ▲Medium | **50A19/31** | 115-125 | .95 | Clear—Commercial Oven | 120 | C-9 | 3 7/8 | 2 9/16 | 1000 | . . . |
| T-10 | Medium | **50T10/RSIF** | 120 | 1.94 | Inside Frosted—Rough Service | 60 | C-8 | 5 5/8 | . . . | 1000 | . . . |
| A-21 | | **50A21** | 12 | .82 | Inside Frosted (11) | 120 | C-6 | 4 7/8 | 3 7/16 | 1000 | 875 |
| ★PAR-36 | Screw Terminal | **50PAR36NSP** | 12 | 3.51 | PAR—Narrow Spotlight (65) | 12 | C-6 | 2 3/4 | . . . | 2000 | . . . |
| | | **50PAR36WFL** | 12 | 3.51 | PAR—Wide Floodlight (65) | 12 | C-6 | 2 3/4 | . . . | 2000 | . . . |
| | | **50PAR36VWFL** | 12 | 3.51 | PAR—Very Wide Floodlight (65) | 12 | C-6 | 2 3/4 | . . . | 2000 | . . . |
| A-21 | Medium | **50A21** | 30 | .59 | Inside Frosted—Train (11) | 120 | C-9 | 4 7/8 | 3 7/16 | 1000 | 805 |
| A-19 | | **50A/RS** | 32 | .70 | Inside Frosted—Rough Service. Train | 120 | C-9 | 3 7/8 | 2 9/16 | 1000 | . . . |
| A-21 | | **50A21** | 34 | .59 | Inside Frosted—Train (11) | 120 | C-9 | 4 7/8 | 3 7/16 | 1000 | . . . |
| | | **50A19/RS** | 75 | .59 | Inside Frosted—Rough Service. Train | 120 | C-9 | 3 7/8 | 2 1/2 | 1000 | 545 |

A-19

| | | | | | | | | | | | |
|---|---|---|---|---|---|---|---|---|---|---|---|
| | | **50A** | **230,250** | **.68** | Inside Frosted (11) | 120 | C-17A | 3 7/8 | 2 9/16 | 1000 | 490 |
| | | **50A/RS** | **230,250** | **.88** | Inside Frosted—Rough Service | 120 | C-22 | 3 7/8 | 2 9/16 | 1000 | 470 |
| | | **50A/G** | **250** | **.81** | Green | 120 | C-17A | 3 7/8 | . . . | 1000 | . . . |
| | | **50A/R** | **250** | **.81** | Red | 120 | C-17A | 3 7/8 | . . . | 1000 | . . . |
| | | **50A19** | **300** | **.70** | Inside Frosted | 120 | C-17A | 3 7/8 | 2 9/16 | 1000 | . . . |
| | | **50A19/RS** | **300** | **.95** | Inside Frosted—Rough Service | 120 | C-22 | 3 7/8 | 2 9/16 | 1000 | . . . |
| | | **50A19/35** | **300** | **.75** | Clear | 120 | C-17A | 3 7/8 | 2 9/16 | 1000 | . . . |

## 50-50 WATTS

| | | | | | | | | | | | |
|---|---|---|---|---|---|---|---|---|---|---|---|
| P-25 | ▲3-Contact Mogul | **50/50P25/28** | **120** | **6.20** | Clear—Marine Running Light. Filament operated separately (1) | 60 | C-5 C-9 | 5 1/16 | 3 5/16 | 750 | 390 440 |
| T-12 | ▲3-Contact Medium | **50/50T12** | **115** | **9.90** | Clear—Marine Running Light. Filament operated separately (1) | 24 | C-5 C-9 | 5 5/16 | 3 | 750 | . . . |

## 50-100-150 WATTS

| | | | | | | | | | | | |
|---|---|---|---|---|---|---|---|---|---|---|---|
| A-21 | 3-Contact Medium | **50/150AX/W 12PK PM** | **120** | **.91** | Soft-White—3-Way. Burn base down. 12-Pack **PRICE MARKED** | 60 | C-2R CC-8 | 5 1/4 | 3 3/4 | 1500, 1200, 1150 | 580, 1640, 2220 |
| A-23 | | **50/150A/W 12PK PM** | **120** | **.91** | Soft-White—3-Way. Burn base down. 12-Pack **PRICE MARKED** (91) | 60 | 2CC-6 | 5 15/16 | 3 7/8 | 1500, 1200, 1150 | 550, 1530, 2080 |
| A-21 | | **50/150AX/DPK 12PK PM** | **115-125** | **1.02** | Coloramic—Dawn Pink. 3-Way. Burn base down. 12-Pack **PRICE MARKED** | 60 | C-2R CC-8 | 5 1/4 | 3 3/4 | 1500, 1200, 1150 | . . . |

## 50-200-250 WATTS

| | | | | | | | | | | | |
|---|---|---|---|---|---|---|---|---|---|---|---|
| PS-25 | 3-Contact Medium | **50/250M/W 12PK PM** | **120** | **1.01** | Soft-White—3-Way. Burn base down. 12-Pack **PRICE MARKED** | 60 | C-2R CC-8 | 6 1/8 | 4 5/8 | 1500, 1200, 1150 | 580, 3660, 4240 |

## 56 WATTS

| | | | | | | | | | | | |
|---|---|---|---|---|---|---|---|---|---|---|---|
| A-21 | Medium | **56A21** | **120** | **.52** | Inside Frosted—Street Railway. Design amps: .519 (41) | 120 | C-9 | 4 3/8 | 2 15/16 | 2000 | . . . |

## INCANDESCENT LAMPS

| Bulb | Base | Lamp Ordering Code | Volts | Approx. List Price | DESCRIPTION | Std. Pkg. Qty. | Filament Design | M. O. L. | L. C. L. | Approx. Hours Life | Approx. Initial Lumens |
|---|---|---|---|---|---|---|---|---|---|---|---|
| **60 WATTS** | | | | | | | | | | | |
| A-19  | Medium | **60A** | 120 | **$0.34** | Inside Frosted (11) | 120 | CC-6 | $4\frac{7}{16}$ | $3\frac{1}{8}$ | 1000 | 870 |
| | | **60A 24PK PM** | 120 | **.34** | Inside Frosted. 24-Pack **PRICE MARKED** | 120 | CC-6 | $4\frac{7}{16}$ | $3\frac{1}{8}$ | 1000 | 870 |
| | | **60A/TF 24PK** | 115-125 | **.82** | Inside Frosted—TUFF-SKIN 24-Pack (44) | 120 | CC-6 | $4\frac{7}{16}$ | ... | 1000 | ... |
| | | **60A** | 125 | **.37** | Inside Frosted (11) | 120 | CC-6 | $4\frac{7}{16}$ | $3\frac{1}{8}$ | 1000 | 870 |
| | | **60A** | 130 | **.42** | Inside Frosted (11) | 120 | CC-6 | $4\frac{7}{16}$ | $3\frac{1}{8}$ | 1000 | 870 |
| | | **60A/W 24PK PM** | 120 | **.37** | Soft-White. 24-Pack **PRICE MARKED** (11) | 120 | CC-6 | $4\frac{7}{16}$ | ... | 1000 | 855 |
| | | **60A/WP 24PK PM** | 120 | **4/1.98** | Soft-White *PLUS*—24-Pack **PRICE MARKED** (11) | 120 | CC-6 | $4\frac{7}{16}$ | $3\frac{1}{8}$ | 1500 | 820• |
| | | **60A/CL** | 120 | **.39** | Clear (11) | 120 | CC-6 | $4\frac{7}{16}$ | $3\frac{1}{8}$ | 1000 | 870 |
| | | **60A/CL 24PK PM** | 120 | **.39** | Clear. 24-Pack **PRICE MARKED** (11) | 120 | CC-6 | $4\frac{7}{16}$ | $3\frac{1}{8}$ | 1000 | 870 |
| | | **60A/CL** | 130 | **.47** | Clear (11) | 120 | CC-6 | $4\frac{7}{16}$ | $3\frac{1}{8}$ | 1000 | 870 |
| | ▲Medium | **60A/99** | 120 | **.47** | Inside Frosted—Extended Service (11) | 120 | CC-6 | $4\frac{7}{16}$ | $3\frac{1}{8}$ | 2500 | 775 |
| | | **60A/99 24PK** | 120 | **.47** | Inside Frosted—Extended Service. 24-Pack (11) | 120 | CC-6 | $4\frac{7}{16}$ | $3\frac{1}{8}$ | 2500 | 775 |
| | | **60A/99** | 130 | **.57** | Inside Frosted—Extended Service (11) | 120 | CC-6 | $4\frac{7}{16}$ | $3\frac{1}{8}$ | 2500 | 775 |
| | Medium | **60A/D** | 115-125 | **.70** | Inside Frosted—Daylight (11) | 120 | CC-6 | $4\frac{7}{16}$ | $3\frac{1}{8}$ | 1000 | 500 |
| | | **60A/DCL** | 115-125 | **.93** | Clear—Daylight (11) | 120 | CC-6 | $4\frac{7}{16}$ | $3\frac{1}{8}$ | 1000 | ... |
| | | **60A/SB** | 120 | **.86** | Inside Frosted—Silvered Bowl (11) | 120 | CC-6 | $4\frac{7}{16}$ | $3\frac{1}{8}$ | 1000 | 740 |
| | | **60A/Y 24PK PM** | 115-125 | **.52** | Yellow—Bug-Lite. 24-Pack **PRICE MARKED** (11) | 120 | CC-6 | $4\frac{7}{16}$ | .. | 1000 | 550 |

| | | | | | | | | | | | | |
|---|---|---|---|---|---|---|---|---|---|---|---|---|
| | | **60A/Y** | **24PK PM** | 130 | .63 | Yellow—Bug-Lite. 24-Pack **PRICE MARKED** (11) | 120 | CC-6 | 4 7/16 | ... | 1000 | 550 |
| | | **60A/DPK** | **24PK PM** | 115-125 | .52 | Coloramic—Dawn Pink. 24-Pack **PRICE MARKED** (11) | 120 | CC-6 | 4 7/16 | ... | 1000 | ... |
| | | **60A/SKY** | **24PK PM** | 115-125 | .52 | Coloramic—Sky Blue. 24-Pack **PRICE MARKED** (11) | 120 | CC-6 | 4 7/16 | ... | 1000 | ... |
| B-10 | ▲Cand. | **60B10/F** | **PM** | 115-125 | .79 | Crystal (Clear) (85) FLAIR | 120 | CC-2V | 4 | ... | 1500 | ... |
| CA-9 | ▲Medium | **60CA9/F** | **PM** | 115-125 | .79 | Clear—Bent Tip (85) FLAIR | 120 | CC-2V | 4 1/2 | 2 5/16 | 1500 | ... |
| | | **60CA9/2/F** | **PM** | 115-125 | .89 | Outside Frosted—Bent Tip (85) FLAIR | 120 | CC-2V | 4 1/2 | 2 5/16 | 1500 | ... |
| CA-10 | ▲Cand. | **60CA10/F** | **PM** | 115-125 | .79 | Clear—Bent Tip (85) FLAIR | 120 | CC-2V | 4 5/16 | 2 1/8 | 1500 | ... |
| | | **60CA10/2/F** | **PM** | 115-125 | .89 | Outside Frosted—Bent Tip (85) FLAIR | 120 | CC-2V | 4 5/16 | 2 1/8 | 1500 | ... |
| A-21  | Medium | **60A21/B** | | 115-125 | .83 | Blue (11,49) | 120 | C-9 | 4 1/8 | ... | 1000 | ... |
| | | **60A21/G** | | 115-125 | .83 | Green (11,49) | 120 | C-9 | 4 1/8 | ... | 1000 | ... |
| | | **60A21/O** | | 115-125 | .83 | Orange (11,49) | 120 | C-9 | 4 1/8 | ... | 1000 | ... |
| | | **60A21/R** | | 115-125 | .83 | Red (11,49) | 120 | C-9 | 4 1/8 | ... | 1000 | ... |
| | | **60A21/R2** | | 115-125 | .83 | Rose (11,49) | 120 | C-9 | 4 1/8 | ... | 1000 | ... |
| | | **60A21/Y** | | 115-125 | .83 | Yellow (11,49) | 120 | C-9 | 4 1/8 | ... | 1000 | ... |
| | | **60A21/NA** | | 115-125 | **6.20** | Natural Amber (11) | 120 | C-9 | 4 1/8 | ... | 1000 | ... |
| | | **60A21/NB** | | 115-125 | **6.20** | Natural Blue (11) | 120 | C-9 | 4 1/8 | ... | 1000 | ... |
| | | **60A21/NG** | | 115-125 | **6.20** | Natural Green (11) | 120 | C-9 | 4 1/8 | ... | 1000 | ... |
| | | **60A21/NR** | | 115-125 | **6.20** | Natural Ruby (11) | 120 | C-9 | 4 1/8 | ... | 1000 | ... |
| T-8  | Disc | **L60** | | 115-125 | **3.07** | Clear—Lumiline | 24 | C-8 | 17 3/4° | ... | 1500 | ... |
| | | **L60/IF** | | 115-125 | **3.21** | Inside Frosted—Lumiline | 24 | C-8 | 17 3/4° | ... | 1500 | 535 |
| | | **L60/W** | | 115-125 | **3.21** | White—Lumiline | 24 | C-8 | 17 3/4° | ... | 1500 | 450 |
| A-21 | ▲Medium | **60A21/TS** | | 120 | .47 | Clear—Traffic Signal (57) | 120 | C-9 | 4 3/8 | 2 7/16 | 3000 | 675 |
| | | **60A21/TS** | | 125 | .51 | Clear—Traffic Signal (57) | 120 | C-9 | 4 3/8 | 2 7/16 | 3000 | 675 |
| | | **60A21/TS** | | 130 | .56 | Clear—Traffic Signal (57) | 120 | C-9 | 4 3/8 | 2 7/16 | 3000 | 675 |

All FLAIR Decorative Lamps are individually packaged, price marked and are available in a 12-pack reshippable specialty pack.

# INCANDESCENT LAMPS

| Bulb | Base | Lamp Ordering Code | Volts | Approx. List Price | DESCRIPTION | Std. Pkg. Qty. | Filament Design | M. O. L. | L. C. L. | Approx. Hours Life | Approx. Initial Lumens |
|---|---|---|---|---|---|---|---|---|---|---|---|
| **60 WATTS (Continued)** | | | | | | | | | | | |
| T-10 | Medium | 60T10/64 24PK | 120 | $1.89 | Clear—Showcase. 24-Pack (57) | 192 | C-8 | 5⅝ | . . . | 1000 | 745 |
| | | 60T10/IF 24PK | 120 | 1.89 | Inside Frosted—Showcase. 24-Pack (57) | 192 | C-8 | 5⅝ | . . . | 1000 | . . . |
| ★PAR-46 | Screw Terminal | 60PAR | 38 | 5.85 | PAR—Mine Locomotive Headlight (71) | 12 | CC-2V | 3¾ | . . . | 800 | . . . |
| | Mogul End Prong | 60PAR/1 | 38 | 5.90 | PAR—Subway Car Headlight (72) | 12 | CC-2V | 4⅜ | . . . | 800 | . . . |
| | Screw Terminal | 60PAR/2/R | 38 | 9.25 | PAR—Red Lens—Train Warning (71) | 12 | CC-2V | 3¾ | . . . | 800 | . . . |
| | | 60PAR/3 | 38 | 5.95 | PAR—Transit Locomotive Headlight (71) | 12 | CC-2V | 3¾ | . . . | 800 | . . . |
| A-21 | Medium | 60A21 | 230 | .70 | Inside Frosted | 120 | C-17A | 4⅜ | 2 15/16 | 1000 | . . . |
| **67 WATTS** | | | | | | | | | | | |
| A-21 | ▲Medium | 67A21/TS | 120 | .49 | Clear—Traffic Signal (57) | 120 | C-9 | 4⅜ | 2 7/16 | 8000 | 635 |
| | | 67A21/TS | 125 | .54 | Clear—Traffic Signal (57) | 120 | C-9 | 4⅜ | 2 7/16 | 8000 | 635 |
| | | 67A21/TS | 130 | .59 | Clear—Traffic Signal (57) | 120 | C-9 | 4⅜ | 2 7/16 | 8000 | 635 |
| **69 WATTS** | | | | | | | | | | | |
| A-21 | ▲Medium | 69A21/TS | 120 | .49 | Clear—Traffic Signal (57) | 120 | C-9 | 4⅜ | 2 7/16 | 8000 | 675 |
| | | 69A21/TS | 125 | .54 | Clear—Traffic Signal (57) | 120 | C-9 | 4⅜ | 2 7/16 | 8000 | 675 |
| | | 69A21/TS | 130 | .59 | Clear—Traffic Signal (57) | 120 | C-9 | 4⅜ | 2 7/16 | 8000 | 675 |
| **75 WATTS** | | | | | | | | | | | |
| A-19 | Medium | 75A | 120 | .34 | Inside Frosted (11) | 120 | CC-6 | 4 7/16 | 3⅛ | 750 | 1190 |
| | | 75A 24PK PM | 120 | .34 | Inside Frosted. 24-Pack **PRICE MARKED** (11) | 120 | CC-6 | 4 7/16 | 3⅛ | 750 | 1190 |

| | | | | | | | | | | | | |
|---|---|---|---|---|---|---|---|---|---|---|---|---|
| | | **75A** | | 125 | .37 | Inside Frosted (11) | 120 | CC-6 | 4 7/16 | 3 1/8 | 750 | 1190 |
| | | **75A** | | 130 | .42 | Inside Frosted (11) | 120 | CC-6 | 4 7/16 | 3 1/8 | 750 | 1190 |
| | | **75A/W** | **24PK PM** | 120 | .37 | Soft-White. 24-Pack. **PRICE MARKED** (11) | 120 | CC-6 | 4 7/16 | 3 1/8 | 750 | 1170 |
| | | **75A/WP** | **24PK PM** | 120 | 4/1.98 | Soft-White *PLUS*. 24-Pack **PRICE MARKED** (11) | 120 | CC-6 | 4 7/16 | 3 1/8 | 1500 | 1075 |
| | | **75A/CL** | | 120 | .39 | Clear (11) | 120 | CC-6 | 4 7/16 | 3 1/8 | 750 | 1190 |
| | | **75A/CL** | | 130 | .47 | Clear (11) | 120 | CC-6 | 4 7/16 | 3 1/8 | 750 | 1190 |
| | ▲Medium | **75A/99** | | 120 | .47 | Inside Frosted—Extended Service (11) | 120 | CC-6 | 4 7/16 | 3 1/8 | 2500 | 1000 |
| | | **75A/99** | **24PK** | 120 | .47 | Inside Frosted—Extended Service. 24-Pack (11) | 120 | CC-6 | 4 7/16 | 3 1/8 | 2500 | 1000 |
| | | **75A/99** | | 130 | .57 | Inside Frosted—Extended Service (11) | 120 | CC-6 | 4 7/16 | 3 1/8 | 2500 | 1000 |
| | Medium | **75A/DPK** | **24PK PM** | 115-125 | .52 | Coloramic—Dawn Pink. 24-Pack **PRICE MARKED** (11) | 120 | CC-6 | 4 7/16 | . . . | 1000 | . . . |
| | | **75A/SKY** | **24PK PM** | 115-125 | .52 | Coloramic—Sky Blue. 24-Pack **PRICE MARKED** (11) | 120 | CC-6 | 4 7/16 | . . . | 1000 | . . . |
| | | **75A/RX** | | 115-125 | .70 | Red—Exit Light (6,11) | 120 | CC-6 | 4 7/16 | 3 1/8 | 5000+ | . . . |
| A-21  | | **75A21/RS** | | 120 | .83 | Inside Frosted—Rough Service | 120 | C-22 | 4 3/8 | 2 11/16 | 1000 | 750 |
| | | **75A21/RS** | **24PK** | 120 | .83 | Inside Frosted—Rough Service. 24-Pack | 120 | C-22 | 4 3/8 | 2 11/16 | 1000 | 750 |
| | ▲Medium | **75A21/RS/TF** | **24PK** | 115-125 | 1.35 | Inside Frosted—Rough Service. TUFF-SKIN. 24-Pack | 120 | C-22 | 4 3/8 | . . . | 1000 | . . . |
| | Medium | **75A21/RS** | | 125-130 | .99 | Inside Frosted—Rough Service | 120 | C-22 | 4 3/8 | 2 11/16 | 1000 | 750 |
| | | **75A21/RS** | **24PK** | 125-130 | .99 | Inside Frosted—Rough Service. 24-Pack | 120 | C-22 | 4 3/8 | 2 11/16 | 1000 | 750 |
| ★PAR-38 | Medium Skirted | **75PAR/FL** | | 120 | 2.59 | PAR—Projector Floodlight | 12 | CC-6 | 5 5/16 | . . | 2000 | 765 |
| | | **75PAR/FL** | **PM** | 120 | 2.59 | PAR—Projector Floodlight. **PRICE MARKED** | 12 | CC-6 | 5 5/16 | . . . | 2000 | 765 |

## INCANDESCENT LAMPS

### 75 WATTS (Continued)

| Bulb | Base | Lamp Ordering Code | Volts | Approx. List Price | DESCRIPTION | Std. Pkg. Qty. | Filament Design | M. O. L. | L. C. L. | Approx. Hours Life | Approx. Initial Lumens |
|---|---|---|---|---|---|---|---|---|---|---|---|
| ★PAR-38 | Medium Skirted | 75PAR/FL | 125-130 | $3.11 | PAR—Projector Floodlight | 12 | CC-6 | 5⅞₆ | ⋯ | 2000 | 765 |
| | | 75PAR/FL  PM | 125-130 | 3.11 | PAR—Projector Floodlight PRICE MARKED | 12 | CC-6 | 5⅞₆ | ⋯ | 2000 | 765 |
| | | 75PAR/SP | 120 | 2.59 | PAR—Projector Spotlight | 12 | CC-6 | 5⅞₆ | ⋯ | 2000 | 765 |
| | | 75PAR/SP  PM | 120 | 2.59 | PAR—Projector Spotlight PRICE MARKED | 12 | CC-6 | 5⅞₆ | ⋯ | 2000 | 765 |
| | | 75PAR/SP | 125-130 | 3.11 | PAR—Projector Spotlight | 12 | CC-6 | 5⅞₆ | ⋯ | 2000 | 765 |
| | | 75PAR/SP  PM | 125-130 | 3.11 | PAR—Projector Spotlight PRICE MARKED | 12 | CC-6 | 5⅞₆ | ⋯ | 2000 | 765 |
| | | 75PAR38/2FL | 120 | 7.00 | PAR—Cool-Beam—Floodlight (14,55,77) | 12 | CC-6 | 5⅞₆ | ⋯ | 2000 | ⋯ |
| | | 75PAR38/2FL | 125-130 | 8.40 | PAR—Cool-Beam—Floodlight (14,55,77) | 12 | CC-6 | 5⅞₆ | ⋯ | 2000 | ⋯ |
| | Medium Side Prong | 75PAR/3FL | 120 | 3.78 | PAR—Compact Projector Flood-light. Support lamp by bulb rim or metal shell of base | 12 | CC-6 | 4⅞₆ | ⋯ | 2000 | 765 |
| | | 75PAR/3SP | 120 | 3.78 | PAR—Compact Projector Spot-light. Support lamp by bulb rim or metal shell of base | 12 | CC-6 | 4⅞₆ | ⋯ | 2000 | 765 |
| R-30 | Medium | 75R30/FL | 120 | 1.51 | Reflector Floodlight (14) | 24 | CC-6 | 5¾ | ⋯ | 2000 | 900 |
| | | 75R30/FL | 125-130 | 1.81 | Reflector Floodlight (14) | 24 | CC-6 | 5¾ | ⋯ | 2000 | 900 |
| | | 75R30/SP | 120 | 1.51 | Reflector Spotlight (14) | 24 | CC-6 | 5¾ | ⋯ | 2000 | 900 |
| | | 75R30/SP | 125-130 | 1.81 | Reflector Spotlight (14) | 24 | CC-6 | 5¾ | ⋯ | 2000 | 900 |
| | | 75R30/A | 115-125 | 2.64 | Reflector—Amber (14,34,35) | 24 | CC-6 | 5¾ | ⋯ | 2000 | ⋯ |
| | | 75R30/B | 115-125 | 2.64 | Reflector—Blue (14,34,35) | 24 | CC-6 | 5¾ | ⋯ | 2000 | ⋯ |
| | | 75R30/BW | 115-125 | 2.64 | Reflector—Blue-White (14,34,35) | 24 | CC-6 | 5¾ | ⋯ | 2000 | ⋯ |

| | | | | | | | | | | | |
|---|---|---|---|---|---|---|---|---|---|---|---|
| | | **75R30/G** | **115-125** | **2.64** | Reflector—Green (14,34,35) | 24 | CC-6 | 5 3/8 | . . . | 2000 | . . . |
| | | **75R30/PK** | **115-125** | **2.64** | Reflector—Pink (14,34,35) | 24 | CC-6 | 5 3/8 | . . . | 2000 | . . . |
| | | **75R30/R** | **115-125** | **2.64** | Reflector—Red (14,34,35) | 24 | CC-6 | 5 3/8 | . . . | 2000 | . . . |
| | | **75R30/Y** | **115-125** | **2.64** | Reflector—Yellow (14,34,35) | 24 | CC-6 | 5 3/8 | . . . | 2000 | . . . |
| R-40 | ▲Medium | **75R/FL** | **120** | **2.07** | Reflector Floodlight—I.F. (34) | 24 | CC-6 | 6 9/16 | . . . | 2000 | . . . |
| | | **75R/FL** | **125-130** | **2.49** | Reflector Floodlight—I.F. (34) | 24 | CC-6 | 6 9/16 | . . . | 2000 | . . . |
| E-17 | | **75E17/W/TF/F PM** | **115-125** | **1.39** | Postlight—Soft-White. TUFF-SKIN (44) FLAIR | 120 | CC-6 | 5 | 3 1/8 | 4000 | . . . |
| G-40 | | **75G40/W/F PM** | **115-125** | **2.25** | White—Moonglow FLAIR | 24 | C-9 | 6 13/16 | . . . | 2500 | 860 |
| | | **75G40/W** | **130** | **2.25** | White—Decorative | 24 | C-9 | 6 13/16 | . . . | 2500 | 860 |
| T-10 | Medium | **75T10/1** | **120** | **2.70** | Inside Frosted—Showcase | 24 | C-23 | 11 7/8 | . . . | 1000 | . . . |
| | | **75T10/45** | **120** | **2.53** | Clear—Showcase | 24 | C-23 | 11 7/8 | . . . | 1000 | . . . |
| A-21 | | **75A21** | **6** | **.91** | Inside Frosted (11,53) | 120 | C-6 | 5 1/4 | 3 13/16 | 1000 | . . . |
| | | **75A21** | **12** | **.91** | Inside Frosted (11,53) | 120 | C-6 | 5 1/4 | 3 13/16 | 1000 | . . . |
| A-23 | | **75A23** | **30** | **.70** | Inside Frosted—Train (11) | 120 | C-9 | 5 15/16 | 4 7/16 | 1000 | . . . |
| ★PAR-38 | Medium Skirted | **75PAR/FL** | **30** | **3.72** | PAR—Train—Floodlight | 12 | C-6 | 5 5/16 | . . . | 1000 | 975 |
| | | **75PAR/FL** | **36** | **3.72** | PAR—Marine Floodlight—Fishing Boat | 12 | C-6 | 5 5/16 | . . . | 1000 | . . . |
| ★PAR-46 | Screw Terminal | **75PAR46** | **48** | **6.15** | PAR—Mine Locomotive Head-light (2,71) | 12 | CC-2V | 3 3/4 | . . . | 800 | . . . |
| ★PAR-38 | Medium Skirted | **75PAR/FL** | **75** | **3.83** | PAR—Train—Floodlight | 12 | CC-6 | 5 5/16 | . . . | 1000 | . . . |
| ★PAR-36 | Screw Terminal | **75PAR36/RS** | **75** | **8.10** | PAR—Train Warning—Rough Service | 12 | CC-6 | 2 3/4 | . . . | 500 | . . . |
| A-21 | Medium | **75A21** | **230** | **.81** | Inside Frosted (11) | 120 | C-7A | 5 1/4 | 3 13/16 | 1000 | . . . |

## 94 WATTS

| | | | | | | | | | | | |
|---|---|---|---|---|---|---|---|---|---|---|---|
| P-25 | Medium | **94P25** | **125** | **2.48** | Clear—Street Railway Head-light. Design amps. .863 (41) | 60 | C-5 | 4 3/4 | 2 1/16 | 1000 | . . . |

All FLAIR Decorative Lamps are individually packaged, price marked, and are available in a 12-lamp reshippable specialty pack.

# INCANDESCENT LAMPS

| Bulb | Base | Lamp Ordering Code | Volts | Approx. List Price | DESCRIPTION | Std. Pkg. Qty. | Filament Design | M.O.L. | L.C.L. | Approx. Hours Life | Approx. Initial Lumens |
|---|---|---|---|---|---|---|---|---|---|---|---|
| **100 WATTS** | | | | | | | | | | | |
| A-19 | Medium | 100A | 120 | $0.34 | Inside Frosted—Bonus Line (46) | 120 | CC-8 | 4 11/16 | 3 5/8 | 750 | 1750 |
| | | 100A 24PK PM | 120 | .34 | Inside Frosted—Bonus Line. 24-Pack. PRICE MARKED (46) | 120 | CC-8 | 4 11/16 | 3 5/8 | 750 | 1750 |
| | | 100A | 125 | .37 | Inside Frosted—Bonus Line (46) | 120 | CC-8 | 4 11/16 | 3 5/8 | 750 | 1750 |
| | | 100A | 130 | .42 | Inside Frosted—Bonus Line (46) | 120 | CC-8 | 4 11/16 | 3 5/8 | 750 | 1750 |
| | | 100A/W 24PK PM | 120 | .37 | Soft-White—Bonus Line. 24-Pack. PRICE MARKED (46) | 120 | CC-8 | 4 11/16 | ... | 750 | 1710 |
| | | 100A/WP 24PK PM | 120 | 4/1.98 | Soft-White PLUS. 24-Pack. PRICE MARKED (11) | 120 | CC-8 | 4 11/16 | ... | 1500 | 1585 |
| | | 100A/CL | 120 | .39 | Clear—Bonus Line (46) | 120 | CC-8 | 4 11/16 | 3 5/8 | 750 | 1750 |
| | | 100A/CL 24PK PM | 120 | .39 | Clear—Bonus Line. 24-Pack. PRICE MARKED (46) | 120 | CC-8 | 4 11/16 | 3 5/8 | 750 | 1750 |
| | | 100A/CL | 130 | .47 | Clear—Bonus Line (46) | 120 | CC-8 | 4 11/16 | 3 5/8 | 750 | 1750 |
| | ▲Medium | 100A/99 | 120 | .50 | Inside Frosted—Extended Service (46) | 120 | CC-8 | 4 11/16 | 3 5/8 | 2500 | 1490 |
| | | 100A/99 24PK | 120 | .50 | Inside Frosted—Extended Service. 24-Pack (46) | 120 | CC-8 | 4 11/16 | 3 5/8 | 2500 | 1490 |
| | | 100A/99 | 130 | .61 | Inside Frosted—Extended Service (46) | 120 | CC-8 | 4 11/16 | 3 5/8 | 2500 | 1490 |
| | ▲Left-Hand | 100A/LHT | 120 | .41 | Inside Frosted—Left-hand threaded base (11) | 120 | CC-8 | 4 11/16 | 3 5/8 | 750 | 1750 |
| | Medium | 100A/LHT | 130 | .49 | Inside Frosted—Left-hand threaded base (11) | 120 | CC-8 | 4 11/16 | 3 5/8 | 750 | 1750 |
| A-21 | Medium | 100A21 | 120 | .40 | Inside Frosted (11) | 120 | CC-6 | 5 1/4 | 3 5/8 | 750 | 1690 |
| | ▲Medium | 100A21/TF | 115-125 | .91 | Inside Frosted—TUFF-SKIN (44) | 120 | CC-6 | 5 1/4 | ... | 750 | ... |

| | | | | | | | | | | | |
|---|---|---|---|---|---|---|---|---|---|---|---|
| | Medium | **100A21** | **125** | **.45** | Inside Frosted (11) | 120 | CC-6 | 5¼ | 3⅞ | 750 | 1690 |
| | | **100A21** | **130** | **.49** | Inside Frosted (11) | 120 | CC-6 | 5¼ | 3⅞ | 750 | 1690 |
| | ▲Medium | **100A21/99** | **120** | **.58** | Inside Frosted—Extended Service (11) | 120 | CC-6 | 5¼ | 3⅞ | 2500 | 1440 |
| | | **100A21/99** | **130** | **.70** | Inside Frosted—Extended Service (11) | 120 | CC-6 | 5¼ | 3⅞ | 2500 | 1440 |
| | Medium | **100A/RS** | **115,120** | **.56** | Inside Frosted—Rough Service (11) | 120 | C-17 | 5¼ | 3 13/16 | 1000 | 1260 |
| | | **100A/RS 24PK** | **120** | **.56** | Inside Frosted—Rough Service. 24-Pack (11) | 120 | C-17 | 5¼ | 3 13/16 | 1000 | 1260 |
| | ▲Medium | **100A/RS/TF** | **115-125** | **1.06** | Inside Frosted—Rough Service. TUFF-SKIN (83) | 120 | C-17 | 5¼ | . . . | 1000 | . . . |
| | Medium | **100A/RS** | **125-130** | **.68** | Inside Frosted—Rough Service (11) | 120 | C-17 | 5¼ | 3 13/16 | 1000 | 1260 |
| | | **100A/RS 24PK** | **125-130** | **.68** | Inside Frosted—Rough Service. 24-Pack (11) | 120 | C-17 | 5¼ | 3 13/16 | 1000 | 1260 |
| | | **100A/CL/RS 24PK** | **120** | **.62** | Clear—Rough Service. 24-Pack (11) | 120 | C-17 | 5¼ | 3 13/16 | 1000 | . . . |
| | | **100A/CL/RS 24PK** | **125-130** | **.75** | Clear—Rough Service. 24-Pack (11) | 120 | C-17 | 5¼ | 3 13/16 | 1000 | . . . |
| A-23 | | **100A23/VS** | **120** | **1.35** | Inside Frosted—Vibration Service (12) | 24 | C-9 | 5 15/16 | 4 7/16 | 1000 | 1340 |
| | | **100A23/VS** | **125-130** | **1.62** | Inside Frosted—Vibration Service (12) | 24 | C-9 | 5 15/16 | 4 7/16 | 1000 | 1340 |
| | | **100A/D** | **115-125** | **.75** | Inside Frosted—Daylight (11) | 120 | CC-6 | 5 15/16 | 4 7/16 | 750 | 910 |
| | | **100A/DCL** | **115-125** | **.86** | Clear—Daylight (11) | 120 | CC-6 | 5 15/16 | 4 7/16 | 750 | . . . |
| A-21 | | **100A/1SBIF** | **120** | **1.12** | Inside Frosted—Silvered Bowl (11,28) | 120 | CC-6 | 5¼ | 3⅞ | 1000 | . . . |
| | | **100A/1SBIF** | **125-130** | **1.35** | Inside Frosted—Silvered Bowl (11,28) | 120 | CC-6 | 5¼ | 3⅞ | 1000 | . . . |
| A-23 | | **100A/SB** | **120** | **1.18** | Inside Frosted—Silvered Bowl (11) | 120 | CC-6 | 5 15/16 | 4 7/16 | 750 | 1470 |
| | | **100A/SB** | **125-130** | **1.42** | Inside Frosted—Silvered Bowl (11) | 120 | CC-6 | 5 15/16 | 4 7/16 | 750 | 1470 |

## INCANDESCENT LAMPS

| Bulb | Base | Lamp Ordering Code | | Volts | Approx. List Price | DESCRIPTION | Std. Pkg. Qty. | Filament Design | M. O. L. | L. C. L. | Approx. Hours Life | Approx. Initial Lumens |
|---|---|---|---|---|---|---|---|---|---|---|---|---|
| **100 WATTS** (Continued) | | | | | | | | | | | | |
| A-21 | Medium | **100A/1Y** | **24PK PM** | **115-125** | **$0.57** | Yellow—Bug-Lite. 24-Pack **PRICE MARKED** (11) | 120 | CC-6 | 5¼ | . . . | 1000 | 1010 |
|  | | **100A/1Y** | **24PK PM** | **130** | **.69** | Yellow—Bug-Lite. 24-Pack **PRICE MARKED** (11) | 120 | CC-6 | 5¼ | . . . | 1000 | 1010 |
| | | **100A/DPK** | **24PK PM** | **115-125** | **.57** | Coloramic—Dawn Pink. 24-Pack **PRICE MARKED** (11) | 120 | CC-6 | 5¼ | . . . | 1000 | . . . |
| | | **100A/SKY** | **24PK PM** | **115-125** | **.57** | Coloramic—Sky Blue. 24-Pack **PRICE MARKED** (11) | 120 | CC-6 | 5¼ | . . . | 1000 | |
| G-40 | ▲Medium | **100G40/CL** | | **115-125** | **2.25** | Clear—Decorative (91) | 24 | C-9 | 6 13/16 | . . . | 2500 | 1300 |
| | | **100G40/W** | | **115-125** | **2.25** | White—Decorative (91) | 24 | C-9 | 6 13/16 | . . . | 2500 | 1160 |
| A-23 | Medium | **100A/B** | | **115-125** | **.95** | Blue (11,49) | 120 | CC-6 | 5 15/16 | . . . | 750 | . . . |
| | | **100A/G** | | **115-125** | **.95** | Green (11,49) | 120 | CC-6 | 5 15/16 | . . . | 750 | . . . |
| | | **100A/O** | | **115-125** | **.95** | Orange (11,49) | 120 | CC-6 | 5 15/16 | . . . | 750 | . . . |
| | | **100A/R** | | **115-125** | **.95** | Red (11,49) | 120 | CC-6 | 5 15/16 | . . . | 750 | . . . |
| A-21 | ▲Medium | **100A21/TS** | | **120** | **.52** | Clear—Traffic Signal (57) | 120 | C-9 | 4⅜ | 2 7/16 | 3000 | 1280 |
|  | | **100A21/TS** | | **125** | **.58** | Clear—Traffic Signal (57) | 120 | C-9 | 4⅜ | 2 7/16 | 3000 | 1280 |
| | | **100A21/TS** | | **130** | **.62** | Clear—Traffic Signal (57) | 120 | C-9 | 4⅜ | 2 7/16 | 3000 | 1280 |
| | | **100A21/SP** | | **120** | **3.21** | Clear—Spotlight (11) | 120 | C-5 | 4⅜ | 3 | 200 | 1340 |
| | | **100A21/SP** | | **125-130** | **3.83** | Clear—Spotlight (11) | 120 | C-5 | 4⅜ | 3 | 200 | 1340 |
| | | **100A21/4SP** | | **120** | **3.78** | Light Inside Frosted—Spotlight (11) | 120 | C-5 | 4⅜ | 3 | 200 | . . . |
| A-21 | Medium Prefocus | **100A21P** | | **120** | **2.43** | Clear—Airport (57) | 120 | CC-2V | 5 5/16 | 2¾ | 2000 | . . . |
| A-23 | Medium | **100A23** | | **115-125** | **.64** | Inside Frosted (11) | 120 | CC-6 | 5 15/16 | 4 7/16 | 750 | . . . |
| | ▲Medium | **100A23/20** | | **115-125** | **.97** | Clear—Commercial Oven (11) | 120 | CC-6 | 5 15/16 | 4 7/16 | 1000 | . . . |
| G-16½ | S.C.Bay. | **100G16½/29SC** | | **120** | **2.91** | Clear—Spotlight (7,57) | 60 | CC-13 | 3 | 1⅜ | 200 | 1660 |

| | | | | | | | | | | | | |
|---|---|---|---|---|---|---|---|---|---|---|---|---|
| | | **100G16½/29SC** | | **125-130** | **3.51** | Clear—Spotlight (7,57) | 60 | CC-13 | 3 | 1 3/8 | 200 | 1660 |
| | D.C.Bay. | **100G16½/29DC** | | **120** | **2.97** | Clear—Spotlight (7,57) | 60 | CC-13 | 3 | 1 3/8 | 200 | 1660 |
| | | **100G16½/29DC** | | **125-130** | **3.56** | Clear—Spotlight (7,57) | 60 | CC-13 | 3 | 1 3/8 | 200 | 1660 |
| ★PAR-38  | Medium Skirted | **100PAR/A** | **PM** | **115-125** | **3.45** | PAR—Projector Flood. Silicone Amber **PRICE MARKED** (14) | 12 | CC-6 | 5 5/16 | . . . | 2000 | . . . |
| | | **100PAR/B** | **PM** | **115-125** | **3.45** | PAR—Projector Flood. Silicone Blue **PRICE MARKED** (14) | 12 | CC-6 | 5 5/16 | . . . | 2000 | . . . |
| | | **100PAR/BW** | **PM** | **115-125** | **3.45** | PAR—Projector Flood. Silicone Blue White **PRICE MARKED** (14) | 12 | CC-6 | 5 5/16 | . . . | 2000 | . . . |
| | | **100PAR/G** | **PM** | **115-125** | **3.45** | PAR—Projector Flood. Silicone Green **PRICE MARKED** (14) | 12 | CC-6 | 5 5/16 | . . . | 2000 | . . . |
| | | **100PAR/PK** | **PM** | **115-125** | **3.45** | PAR—Projector Flood. Silicone Pink **PRICE MARKED** (14) | 12 | CC-6 | 5 5/16 | . . . | 2000 | . . . |
| | | **100PAR/R** | **PM** | **115-125** | **3.45** | PAR—Projector Flood. Silicone Red **PRICE MARKED** (14) | 12 | CC-6 | 5 5/16 | . . . | 2000 | . . . |
| | | **100PAR/Y** | **PM** | **115-125** | **3.45** | PAR—Projector Flood. Silicone Yellow **PRICE MARKED** (14) | 12 | CC-6 | 5 5/16 | . . . | 2000 | . . . |
| R-40 | ▲Medium | **100R/FL** | | **120** | **1.67** | Reflector Floodlight—I.F. (34) | 24 | CC-6 | 6 9/16 | . . . | 2000 | . . . |
| | | **100R/FL** | | **125-130** | **2.00** | Reflector Floodlight—I.F. (34) | 24 | CC-6 | 6 9/16 | . . . | 2000 | . . . |
| | | **100R/SP** | | **120** | **1.67** | Reflector Spotlight—Light Inside Frosted (34) | 24 | CC-6 | 6 9/16 | . . . | 2000 | . . . |
| T-8½ | Medium Prefocus | **100T8½/8** | | **120** | **7.55** | Clear—Microscope. Burn base down. ANSI: CLD (8) | 24 | CC-13 | 5 5/8 | 2 3/16 | 50 | 1920 |
| | ▲Medium | **100T8½/9** | | **120** | **7.00** | Clear—Microscope. Burn base down. ANSI: EDR (8) | 24 | CC-13 | 5 3/8 | 3 | 50 | 1920 |
| ★T-10  | D.C. Medium Ring | **100T10/7** | | **6** | **11.85** | Clear—Contour Projection. Filament offset .100″ ±.030″ from base axis. ANSI: CPS (1) | 24 | C-6 | 5½ | 2 3/16 | 50 | . . . |
| | Medium Prefocus | **100T10P** | | **6** | **9.15** | Clear—Contour Projection. Filament offset .100″ ±.030″ from base axis. ANSI: CPT (1) | 24 | C-6 | 5¾ | 2 3/16 | 50 | . . . |
| A-23 | Medium | **100A23** | | **12** | **.97** | Inside Frosted (11,53) | 120 | C-6 | 5 15/16 | 4 7/16 | 1000 | . . . |
| ★PAR-38 | Med.Side Prong | **100PAR38/FL** | | **12** | **6.65** | PAR—Mine Floodlight (58) | 12 | C-6 | 4 5/16 | . . . | 1000 | . . . |

## INCANDESCENT LAMPS

| Bulb | Base | Lamp Ordering Code | Volts | Approx. List Price | DESCRIPTION | Std. Pkg. Qty. | Filament Design | M. O. L. | L. C. L. | Approx. Hours Life | Approx. Initial Lumens |
|---|---|---|---|---|---|---|---|---|---|---|---|
| **100 WATTS** (Continued) | | | | | | | | | | | |
| ★PAR-38 | Medium Skirted | **100PAR38/2FL** | **12** | **$6.65** | PAR—Projector Floodlight | 12 | C-6 | 5 5/16 | ... | 1000 | ... |
| T-8 | S.C.Bay. | **100T8/1SC** | **20** | **3.51** | Clear—Contour Map. Source WxH: 4.5x3.0mm. ANSI: BZA (3,8,31,61) | 24 | CC-6 | 3 | 2 3/16 | 50 | ... |
| A-21 | Medium | **100A/RS** | **30** | **1.12** | Inside Frosted—Rough Service (11) | 120 | C-9 | 5 1/4 | 3 13/16 | 1000 | ... |
| | ▲Medium | **100A21/3** | **32** | **2.97** | Clear—Locomotive Headlight (13) | 120 | C-5 | 4 3/8 | 3 | 500 | 1610 |
| A-23 | Medium | **100A** | **34** | **.75** | Inside Frosted—Train (11) | 120 | C-9 | 5 15/16 | 4 7/16 | 1000 | 2160 |
| ★PAR-46 | Screw Terminal | **100PAR46** | **60** | **6.75** | PAR—Mine Locomotive Headlight (71) | 12 | CC-2V | 3 3/4 | ... | 800 | ... |
| A-21  | Medium | **100A** | **230,250** | **.66** | Inside Frosted (11) | 120 | C-7A | 5 1/4 | 3 13/16 | 1000 | 1300 |
| | ▲Medium | **100A/TF** | **230-250** | **1.56** | Inside Frosted—TUFF-SKIN (44) | 120 | C-7A | 5 1/4 | ... | 1000 | ... |
| | Medium | **100A/RS** | **250** | **.75** | Inside Frosted—Rough Service (11) | 120 | C-17 | 5 1/4 | 3 13/16 | 1000 | 1030 |
| | ▲Medium | **100A/99** | **230-250** | **.72** | Inside Frosted—Extended Service (11) | 120 | C-7A | 5 1/4 | 3 13/16 | 2500 | ... |
| | Medium | **100A** | **277** | **.75** | Inside Frosted (11) | 120 | C-7A | 5 1/4 | 3 13/16 | 1000 | ... |
| | | **100A/RS** | **300** | **1.18** | Inside Frosted—Rough Service (11) | 120 | C-17 | 5 1/4 | 3 13/16 | 1000 | ... |
| **100-100 WATTS** | | | | | | | | | | | |
| P-25 | ▲3-Contact Mogul | **100/100P25/29** | **120** | **5.80** | Clear—Marine Running Light. Filaments operated separately (1) | 60 | C-5 / C-9 | 5 1/16 | 3 5/16 | 750 | 1080, 1160 |

**100-200-300 WATTS**

| Watts / Type | Base | Ordering Abbrev. | Pkg. | Volts | Price | Description | Pkg. Qty. | Bulb | MOL | LCL | Init. Lumens | Avg. Life |
|---|---|---|---|---|---|---|---|---|---|---|---|---|
| PS-25 | 3-Contact Mogul | 100/300 | 6PK PM | 120 | 1.39 | Soft-White—3-Way. Burn base down. 6-Pack **PRICE MARKED** | 24 | {C-2R/CC-8} | 6 1/16 | ... | 1500, 1200, 1150 | 1320, 3620, 4940 |
| | | 100/300/DPK | 12PK PM | 115-125 | 1.69 | Coloramic—Dawn Pink. 3-Way. Burn base down. 12-Pack **PRICE MARKED** | 60 | {C-2R/CC-8} | 6 1/16 | ... | 1200 | ... |
| | | 100/300/2 | | 120 | 2.21 | Soft-White—3-Way. Neck of bulb coated red from base to maximum bulb diameter. Burn base down | 60 | {C-2R/CC-8} | 6 1/16 | ... | 1200 | ... |
| | | 100/300/3 | | 120 | 2.21 | Soft-White—3-Way. Neck of bulb coated old rose from base to maximum bulb diameter. Burn base down | 60 | {C-2R/CC-8} | 6 1/16 | ... | 1200 | ... |

**101 WATTS**

| Type | Base | Ordering Abbrev. | Volts | Price | Description | Pkg. Qty. | Bulb | MOL | LCL | Init. Lumens | Avg. Life |
|---|---|---|---|---|---|---|---|---|---|---|---|
| A-23 | Medium | 101A23 | 120 | 1.12 | Inside Frosted—Street Railway. Arc Resisting (11.25) | 120 | C-9 | 5 9/16 | 4 7/16 | 1500 | ... |

**110 WATTS**

| Type | Base | Ordering Abbrev. | Volts | Price | Description | Pkg. Qty. | Bulb | MOL | LCL | Init. Lumens | Avg. Life |
|---|---|---|---|---|---|---|---|---|---|---|---|
| R-30 | ▲Medium | 110R30/FL/RS | 120 | 2.48 | Reflector Floodlight—Inside Frosted. Rough Service (14) | 24 | C-17 | 5 3/8 | ... | 2000 | ... |

**116 WATTS**

| Type | Base | Ordering Abbrev. | Volts | Price | Description | Pkg. Qty. | Bulb | MOL | LCL | Init. Lumens | Avg. Life |
|---|---|---|---|---|---|---|---|---|---|---|---|
| A-21 | ▲Medium | 116A21/TS | 120 | .56 | Clear—Traffic Signal. Rated watts: 114 (57) | 120 | C-9 | 4 3/8 | 2 5/16 | 8000 | 1280 |
| | | 116A21/TS | 125 | .61 | Clear—Traffic Signal. Rated watts: 114 (57) | 120 | C-9 | 4 3/8 | 2 5/16 | 8000 | 1280 |
| | | 116A21/TS | 130 | .68 | Clear—Traffic Signal. Rated watts: 114 (57) | 120 | C-9 | 4 3/8 | 2 5/16 | 8000 | 1280 |

**120 WATTS**

| Type | Base | Ordering Abbrev. | Volts | Price | Description | Pkg. Qty. | Bulb | MOL | LCL | Init. Lumens | Avg. Life |
|---|---|---|---|---|---|---|---|---|---|---|---|
| ★PAR-64 | Screw Terminal | 120PAR | 6 | 13.95 | PAR—Transmissometer—Narrow Spotlight | ... | C-6 | 4 | ... | 3000 | ... |
| | | 120PAR/1 | 6 | 13.95 | PAR—Narrow Spotlight | ... | C-6 | 4 | ... | 2000 | ... |
| ★PAR-56 | | 120PAR56VNSP | 12 | 5.55 | PAR—Very Narrow Spotlight | 12 | C-6 | 4 1/2 | ... | 2000 | ... |
| | | 120PAR56MFL | 12 | 5.55 | PAR—Medium Floodlight | 12 | C-6 | 4 1/2 | ... | 2000 | ... |
| | | 120PAR56WFL | 12 | 5.55 | PAR—Wide Floodlight | 12 | C-6 | 4 1/2 | ... | 2000 | ... |

# INCANDESCENT LAMPS

| Bulb | Base | Lamp Ordering Code | Volts | Approx. List Price | DESCRIPTION | Std. Pkg. Qty. | Filament Design | M. O. L. | L. C. L. | Approx. Hours Life | Approx. Initial Lumens |
|---|---|---|---|---|---|---|---|---|---|---|---|
| **125 WATTS** | | | | | | | | | | | |
| G-16½ | D.C.Bay. | **125G16½DC** | 120 | $3.67 | Clear—Spotlight (7,57) | 60 | CC-13B | 3 | 1½ | 600 | . . . |
| | S.C.Bay. | **125G16½/1SC** | 120 | 3.51 | Clear—Spotlight (7,57) | 60 | CC-13B | 3 | 1⅜ | 600 | . . . |
| R-40 | Medium Skirted | **125R40** | 115-125 | 2.53 | Reflector Infrared—Industrial (6,16,34) | 24 | C-9 | 7⅜ | . . . | 5000+ | . . . |
| T-10 | Medium Prefocus | **125T10P** | 120 | 6.20 | Clear—Spotlight. Burn base down | 24 | C-13B | 5¾ | 2 3/16 | 500 | 1820 |
| **150 WATTS** | | | | | | | | | | | |
| A-21  | Medium | **150A** | 120 | .47 | Inside Frosted—Bonus Line (46) | 60 | CC-8 | 5½ | 4 | 750 | 2880 |
| | | **150A 24PK PM** | 120 | .47 | Inside Frosted—Bonus Line. 24-Pack **PRICE MARKED** (46) | 120 | CC-8 | 5½ | 4 | 750 | 2880 |
| | | **150A** | 125 | .51 | Inside Frosted—Bonus Line (46) | 60 | CC-8 | 5½ | 4 | 750 | 2880 |
| | | **150A** | 130 | .57 | Inside Frosted—Bonus Line (46) | 60 | CC-8 | 5½ | 4 | 750 | 2880 |
| | | **150A/W 24PK PM** | 120 | .52 | Soft-White—Bonus Line. 24-Pack **PRICE MARKED** (46) | 120 | CC-8 | 5½ | 4 | 750 | 2790 |
| | | **150A/CL** | 120 | .49 | Clear—Bonus Line (46) | 60 | CC-8 | 5½ | 4 | 750 | 2880 |
| | | **150A/CL** | 130 | .59 | Clear—Bonus Line (46) | 60 | CC-8 | 5½ | 4 | 750 | 2880 |
| A-23  | | **150A23** | 120 | .52 | Inside Frosted (11) | 60 | CC-6 | 6 3/16 | 4⅝ | 750 | 2780 |
| | | **150A23** | 125 | .57 | Inside Frosted (11) | 60 | CC-6 | 6 3/16 | 4⅝ | 750 | 2780 |
| | | **150A23** | 130 | .63 | Inside Frosted (11) | 60 | CC-6 | 6 3/16 | 4⅝ | 750 | 2780 |
| | | **150A23/CL** | 120 | .59 | Clear (11) | 60 | CC-6 | 6 3/16 | 4⅝ | 750 | 2780 |
| | | **150A23/CL** | 130 | .71 | Clear (11) | 60 | CC-6 | 6 3/16 | 4⅝ | 750 | 2780 |
| | ▲Medium | **150A23/99** | 120 | .66 | Inside Frosted—Extended Service (11) | 60 | CC-6 | 6 3/16 | 4⅝ | 2500 | 2310 |
| | | **150A23/99 24PK** | 120 | .66 | Inside Frosted—Extended Service. 24-Pack (11) | 72 | CC-6 | 6 3/16 | 4⅝ | 2500 | 2310 |

| | | | | | | | | | | | |
|---|---|---|---|---|---|---|---|---|---|---|---|
| | | **150A23/99** | **130** | **.79** | Inside Frosted—Extended Service (11) | 60 | CC-6 | 6 3/16 | 4 5/8 | 2500 | 2310 |
| | | **150A23/99CL** | **120** | **.72** | Clear—Extended Service (11) | 60 | CC-6 | 6 3/16 | 4 5/8 | 2500 | 2310 |
| | | **150A23/99CL** | **130** | **.86** | Clear—Extended Service (11) | 60 | CC-6 | 6 3/16 | 4 5/8 | 2500 | 2310 |
| | Medium | **150A/RS** | **120** | **1.12** | Inside Frosted—Rough Service (11) | 24 | C-17 | 5 15/16 | 4 7/16 | 1000 | . . . |
| | | **150A/RS** | **125-130** | **1.33** | Inside Frosted—Rough Service (11) | 24 | C-17 | 5 15/16 | 4 7/16 | 1000 | . . . |
| PS-25  | | **150** | **120** | **.64** | Inside Frosted (11) | 60 | C-9 | 6 15/16 | 5 3/16 | 750 | 2680 |
| | ▲Medium | **150PS25/IF/TF** | **115-125** | **1.45** | Inside Frosted—TUFF-SKIN (44) | 60 | C-9 | 6 15/16 | . . . | 1000 | . . . |
| | Medium | **150** | **125** | **.71** | Inside Frosted (11) | 60 | C-9 | 6 15/16 | 5 3/16 | 750 | 2680 |
| | | **150** | **130** | **.77** | Inside Frosted (11) | 60 | C-9 | 6 15/16 | 5 3/16 | 750 | 2680 |
| | | **150/CL** | **120** | **.71** | Clear (11) | 60 | C-9 | 6 15/16 | 5 3/16 | 750 | 2680 |
| | | **150/CL** | **130** | **.85** | Clear (11) | 60 | C-9 | 6 15/16 | 5 3/16 | 750 | 2680 |
| | ▲Medium | **150/99** | **120** | **.68** | Inside Frosted—Extended Service (11) | 60 | C-9 | 6 15/16 | 5 3/16 | 2500 | 2300 |
| | | **150/99 24PK** | **120** | **.68** | Inside Frosted—Extended Service. 24-Pack (11) | 120 | C-9 | 6 15/16 | 5 3/16 | 2500 | 2300 |
| | | **150/99** | **130** | **.82** | Inside Frosted—Extended Service (11) | 60 | C-9 | 6 15/16 | 5 3/16 | 2500 | 2300 |
| | | **150/99CL** | **120** | **.71** | Clear—Extended Service (11) | 60 | C-9 | 6 15/16 | 5 3/16 | 2500 | 2300 |
| | | **150/99CL** | **130** | **.85** | Clear—Extended Service (11) | 60 | C-9 | 6 15/16 | 5 3/16 | 2500 | 2300 |
| | Medium | **150/RS** | **120** | **1.18** | Inside Frosted—Rough Service (11) | 60 | C-17 | 6 15/16 | 5 3/16 | 1000 | 2160 |
| | | **150/RS 24PK** | **120** | **1.18** | Inside Frosted—Rough Service. 24-Pack (11) | 72 | C-17 | 6 15/16 | 5 3/16 | 1000 | 2160 |
| | ▲Medium | **150/RS/TF** | **115-125** | **1.99** | Inside Frosted—Rough Service. TUFF-SKIN (83) | 60 | C-17 | 6 15/16 | . . . | 1000 | . . . |
| | Medium | **150/RS** | **125-130** | **1.42** | Inside Frosted—Rough Service (11) | 60 | C-17 | 6 15/16 | 5 3/16 | 1000 | 2160 |
| | | **150/RS 24PK** | **125-130** | **1.42** | Inside Frosted—Rough Service. 24-Pack (11) | 72 | C-17 | 6 15/16 | 5 3/16 | 1000 | 2160 |

| Bulb | Base | Lamp Ordering Code | Approx. List Price | Volts | DESCRIPTION | Std. Pkg. Qty. | Filament Design | M.O.L. | L.C.L. | Approx. Hours Life | Approx. Initial Lumens |
|---|---|---|---|---|---|---|---|---|---|---|---|
| **150 WATTS (Continued)** | | | | | | | | | | | |
| PS-25 | Medium | 150/CL/RS | $1.18 | 120 | Clear—Rough Service (11) | 60 | C-17 | 6¹¹⁄₁₆ | 5⁵⁄₁₆ | 1000 | ... |
| | | 150/CL/RS | 1.42 | 125-130 | Clear—Rough Service (11) | 60 | C-17 | 6¹¹⁄₁₆ | 5⁵⁄₁₆ | 1000 | ... |
| | | 150/VS | 1.45 | 120 | Inside Frosted—Vibration Service (11) | 24 | C-9 | 6¹¹⁄₁₆ | 5⁵⁄₁₆ | 1000 | 2400 |
| | | 150/VS | 1.74 | 125-130 | Inside Frosted—Vibration Service (11) | 24 | C-9 | 6¹¹⁄₁₆ | 5⁵⁄₁₆ | 1000 | 2400 |
| | | 150/CL/VS | 1.45 | 120 | Clear—Vibration Service (11) | 60 | C-9 | 6¹¹⁄₁₆ | 5⁵⁄₁₆ | 1000 | ... |
| | | 150/CL/VS | 1.74 | 125-130 | Clear—Vibration Service (11) | 60 | C-9 | 6¹¹⁄₁₆ | 5⁵⁄₁₆ | 1000 | ... |
| | | 150PS25/Y | .91 | 115-125 | Yellow—Bug-Lite (11) | 60 | C-9 | 6¹¹⁄₁₆ | 5⁵⁄₁₆ | 1000 | 1560 |
| | | 150/D | 1.18 | 115-125 | Inside Frosted—Daylight (11) | 60 | C-9 | 6¹¹⁄₁₆ | 5⁵⁄₁₆ | 1000 | ... |
| | | 150/DCL | 1.18 | 115-125 | Clear—Daylight (11) | 60 | C-9 | 6¹¹⁄₁₆ | 5⁵⁄₁₆ | 1000 | ... |
| | | 150/SB | 1.18 | 120 | Inside Frosted—Silvered Bowl (11,14) | 60 | C-9 | 6¹¹⁄₁₆ | 5⁵⁄₁₆ | 1000 | 2370 |
| | | 150/SB | 1.42 | 125-130 | Inside Frosted—Silvered Bowl (11,14) | 60 | C-9 | 6¹¹⁄₁₆ | 5⁵⁄₁₆ | 1000 | 2370 |
| ★G-16½ | S.C.Bay. | 150G16½SC | ♦ | 120 | Clear—Spotlight (7,57) | 60 | 2CC-8 | 3 | 1⅞ | 200 | ... |
| ★PAR-38 | Medium Skirted | 150PAR/FL | 2.60 | 120 | PAR—Clear. Projector Floodlight (14) | 12 | CC-6 | 5⁵⁄₁₆ | ... | 2000 | 1740 |
| | | 150PAR/FL PM | 2.60 | 120 | PAR—Clear—Projector Floodlight PRICE MARKED (14) | 12 | CC-6 | 5⁵⁄₁₆ | ... | 2000 | 1740 |
| | | 150PAR/FL | 3.12 | 125-130 | PAR—Clear—Projector Floodlight (14) | 12 | CC-6 | 5⁵⁄₁₆ | ... | 2000 | 1740 |
| | | 150PAR/FL PM | 3.12 | 125-130 | PAR—Clear—Projector Floodlight PRICE MARKED (14) | 12 | CC-6 | 5⁵⁄₁₆ | ... | 2000 | 1740 |
| | | 150PAR/WFL | 2.60 | 120 | PAR—Projector Wide Floodlight (14) | 12 | CC-6 | 5⁵⁄₁₆ | ... | 2000 | ... |
| | | 150PAR/WFL | 3.12 | 125-130 | PAR—Projector Wide Floodlight (14) | 12 | CC-6 | 5⁵⁄₁₆ | ... | 2000 | ... |

| | | | | | | | | | | | |
|---|---|---|---|---|---|---|---|---|---|---|---|
| | 150PAR38/2FL | | 120 | 7.00 | PAR—Cool-Beam—Floodlight (14,55,77) | 12 | CC-6 | 5⁵⁄₁₆ | ... | 2000 | 1740 |
| | 150PAR38/2FL | | 125-130 | 8.40 | PAR—Cool-Beam—Floodlight (14,55,77) | 12 | CC-6 | 5⁵⁄₁₆ | ... | 2000 | 1740 |
| Medium Side Prong | 150PAR/3FL | | 120 | 3.45 | PAR—Compact Projector Floodlight (58) | 12 | CC-6 | 4⁵⁄₁₆ | ... | 2000 | 1740 |
| | 150PAR/3FL | | 125-130 | 4.15 | PAR—Compact Projector Floodlight (58) | 12 | CC-6 | 4⁵⁄₁₆ | ... | 2000 | 1740 |
| Medium Skirted | 150PAR/SP | | 120 | 2.60 | PAR—Clear—Projector Spotlight (14) | 12 | CC-6 | 5⁵⁄₁₆ | ... | 2000 | 1740 |
| | 150PAR/SP | PM | 120 | 2.60 | PAR—Clear—Projector Spotlight **PRICE MARKED** (14) | 12 | CC-6 | 5⁵⁄₁₆ | ... | 2000 | 1740 |
| | 150PAR/SP | | 125-130 | 3.12 | PAR—Clear—Projector Spotlight (14) | 12 | CC-6 | 5⁵⁄₁₆ | ... | 2000 | 1740 |
| | 150PAR/SP | PM | 125-130 | 3.12 | PAR—Clear—Projector Spotlight **PRICE MARKED** (14) | 12 | CC-6 | 5⁵⁄₁₆ | ... | 2000 | 1740 |
| | 150PAR38/2SP | | 120 | 7.00 | PAR—Cool-Beam—Spotlight (14,55,77) | 12 | CC-6 | 5⁵⁄₁₆ | ... | 2000 | 1740 |
| | 150PAR38/2SP | | 125-130 | 8.40 | PAR—Cool-Beam—Spotlight (14,55,77) | 12 | CC-6 | 5⁵⁄₁₆ | ... | 2000 | 1740 |
| Medium Side Prong | 150PAR/3SP | | 120 | 3.45 | PAR—Compact Projector Spotlight (58) | 12 | CC-6 | 4⁵⁄₁₆ | ... | 2000 | 1740 |
| | 150PAR/3SP | | 125-130 | 4.15 | PAR—Compact Projector Spotlight (58) | 12 | CC-6 | 4⁵⁄₁₆ | ... | 2000 | 1740 |
| Medium Skirted | 150PAR/FL/A | | 115-125 | 8.55 | PAR—Projector Floodlight—Dichro Amber (14,79) | 12 | CC-6 | 5⁵⁄₁₆ | ... | 2000 | ... |
| | 150PAR/FL/B | | 115-125 | 8.55 | PAR—Projector Floodlight—Dichro Blue (14,79) | 12 | CC-6 | 5⁵⁄₁₆ | ... | 2000 | ... |
| | 150PAR/FL/G | | 115-125 | 8.55 | PAR—Projector Floodlight—Dichro Green (14,79) | 12 | CC-6 | 5⁵⁄₁₆ | ... | 2000 | ... |

◆ Now classified as Stage/Studio Lamp type. For information on the complete line of Stage/Studio Lamps, consult your General Electric Stage/Studio Lamp supplier or the nearest GE Lamp Sales District Office or Distribution Center.

# INCANDESCENT LAMPS

## 150 WATTS (Continued)

| Bulb | Base | Lamp Ordering Code | Volts | Approx. List Price | DESCRIPTION | Std. Pkg. Qty. | Filament Design | M. O. L. | L. C. L. | Approx. Hours Life | Approx. Initial Lumens |
|---|---|---|---|---|---|---|---|---|---|---|---|
| ★PAR-38 | Medium Skirted | 150PAR/FL/R | 115-125 | $8.55 | PAR—Projector Floodlight—Dichro Red (14.79) | 12 | CC-6 | 5⁵⁄₁₆ | ... | 2000 | ... |
| | | 150PAR/FL/Y | 115-125 | 8.55 | PAR—Projector Floodlight—Dichro Yellow (14.79) | 12 | CC-6 | 5⁵⁄₁₆ | ... | 2000 | ... |
| | | 150PAR/SP/A | 115-125 | 8.55 | PAR—Projector Spotlight—Dichro Amber (14.79) | 12 | CC-6 | 5⁵⁄₁₆ | ... | 2000 | ... |
| | | 150PAR/SP/B | 115-125 | 8.55 | PAR—Projector Spotlight—Dichro Blue (14.79) | 12 | CC-6 | 5⁵⁄₁₆ | ... | 2000 | ... |
| | | 150PAR/SP/G | 115-125 | 8.55 | PAR—Projector Spotlight—Dichro Green (14.79) | 12 | CC-6 | 5⁵⁄₁₆ | ... | 2000 | ... |
| | | 150PAR/SP/R | 115-125 | 8.55 | PAR—Projector Spotlight—Dichro Red (14.79) | 12 | CC-6 | 5⁵⁄₁₆ | ... | 2000 | ... |
| | | 150PAR/SP/Y | 115-125 | 8.55 | PAR—Projector Spotlight—Dichro Yellow (14.79) | 12 | CC-6 | 5⁵⁄₁₆ | ... | 2000 | ... |
| ★PAR-46 | Screw Terminal | 150PAR46 | 120-130 | 8.20 | PAR—Mine Locomotive Head-light (71) | 12 | C-13 | 3¾ | ... | 1000 | ... |
| ★PAR-38 | Medium Side Prong | 150PAR/4 | 120-130 | 6.75 | PAR—Clear—Mine. Burning position: prongs up or down (58) | 12 | C-13 | 4⁵⁄₁₆ | ... | 1000 | ... |
| | Medium Skirted | 150PAR/5 | 120-130 | 6.40 | PAR—Clear—Mine. Burning position: locating lug up or down. Support lamp by bulb rim (14) | 12 | C-13 | 5⁵⁄₁₆ | ... | 1000 | ... |
| ★PAR-46 | Screw Terminal | 150PAR46/3 | 175 | 9.10 | PAR—Mine Locomotive Head-light (71) | 12 | C-13 | 3¾ | ... | 800 | ... |
| R-40 | ▲Medium | 150R/FL | 120 | 1.60 | Reflector Floodlight—Inside Frosted. ANSI: DWC (22.35) | 24 | CC-6 | 6⁵⁄₁₆ | ... | 2000 | 1870 |
| | | 150R/FL | 125-130 | 1.94 | Reflector Floodlight:Inside Frosted (22.35) | 24 | CC-6 | 6⁵⁄₁₆ | ... | 2000 | 1870 |

| Type | Ordering Abbreviation | Volts | List Price | Description | MSCP | Filament | Max. Length (in.) | MSCP | Rated Life (hrs.) | Approx. Lumens |
|---|---|---|---|---|---|---|---|---|---|---|
| ★R-40 | 150R/3FL | 120 | 4.32 | Reflector Floodlight—I.F. (21) | 24 | CC-6 | 6-3/4 | ... | 2000 | ... |
| R-40 | 150R/3FL | 125-130 | 5.15 | Reflector Floodlight—I.F. (21) | 24 | CC-6 | 6-3/4 | ... | 2000 | 1870 |
| | 150R/SP | 120 | 1.60 | Reflector Spotlight—Light I.F. (22,35) | 24 | CC-6 | 6-9/16 | ... | 2000 | 1870 |
| | 150R/SP | 125-130 | 1.94 | Reflector Spotlight—Light I.F. (22,35) | 24 | CC-6 | 6-9/16 | ... | 2000 | 1870 |
| | 150R/A | 115-125 | 2.90 | Reflector—Amber (22,35) | 24 | CC-6 | 6-9/16 | ... | 2000 | ... |
| | 150R/B | 115-125 | 2.90 | Reflector—Blue (22,35) | 24 | CC-6 | 6-9/16 | ... | 2000 | ... |
| | 150R/BW | 115-125 | 2.90 | Reflector—Blue White (22,35) | 24 | CC-6 | 6-9/16 | ... | 2000 | ... |
| | 150R/G | 115-125 | 2.90 | Reflector—Green (22,35) | 24 | CC-6 | 6-9/16 | ... | 2000 | ... |
| | 150R/PK | 115-125 | 2.90 | Reflector—Pink (22,35) | 24 | CC-6 | 6-9/16 | ... | 2000 | ... |
| | 150R/R | 115-125 | 2.90 | Reflector—Red (22,35) | 24 | CC-6 | 6-9/16 | ... | 2000 | ... |
| | 150R/Y | 115-125 | 2.90 | Reflector—Yellow (22,35) | 24 | CC-6 | 6-9/16 | ... | 2000 | ... |
| P-25 | 150P25/2 | 120 | 3.56 | Clear—Spotlight. Hard glass button | 60 | C-5 | 4-3/4 | 3 | 200 | ... |
| | 150P25/2SB | 120 | 4.10 | Clear—Spotlight. Silvered Bowl. Hard glass button | 60 | C-5 | 4-3/4 | ... | 200 | ... |
| | 150P25/8 | 120 | 3.78 | Medium Inside Frosted—Spotlight. Hard glass button | 60 | C-5 | 4-3/4 | 3 | 200 | ... |
| | 150P25/8SB | 120 | 4.64 | Medium Inside Frosted—Spotlight. Silvered Bowl. Hard glass button | 60 | C-5 | 4-3/4 | ... | 200 | ... |
| | 150P25/10 | 120 | 3.78 | Light Inside Frosted—Spotlight. Hard glass button | 60 | C-5 | 4-3/4 | 3 | 200 | 2100 |
| | 150P25/10 | 125-130 | 4.59 | Light Inside Frosted—Spotlight. Hard glass button | 60 | C-5 | 4-3/4 | 3 | 200 | 2100 |
| | 150P25/10SB | 120 | 4.64 | Light Inside Frosted—Spotlight. Silvered Bowl. Hard glass button | 60 | C-5 | 4-3/4 | ... | 200 | ... |
| | 150P25/15 | 120 | 3.40 | Clear—Train Headlight (57) | 60 | C-5 | 4-3/4 | 3 | 500 | 1920 |
| PS-30 | 150PS30/1 | 120 | 1.35 | Clear—Silvered Neck | 60 | C-9 | 8-1/16 | 6 | 1000 | ... |
| ★PAR-38 Medium Side Prong | 150PAR38 | 32 | 8.60 | PAR—Clear—Mine. Burning position: prongs up or down (58) | 12 | CC-8 | 4-5/16 | ... | 1000 | ... |

## INCANDESCENT LAMPS

| Bulb | Base | Lamp Ordering Code | Volts | Approx. List Price | DESCRIPTION | Std. Pkg. Qty. | Filament Design | M. O. L. | L. C. L. | Approx. Hours Life | Approx. Initial Lumens |
|---|---|---|---|---|---|---|---|---|---|---|---|
| **150 WATTS (Continued)** | | | | | | | | | | | |
| ★PAR-46 | Screw Terminal | 150PAR46/1 | 32 | $7.25 | PAR—Mine Locomotive Head-light | 12 | CC-8 | 3¾ | ... | 800 | ... |
| PS-25 | Medium | 150PS25/IF | 230,250 | 1.06 | Inside Frosted (11) | 60 | C-7A | 6 15/16 | 5 3/16 | 1000 | ... |
| | | 150PS25 | 230 | 1.02 | Clear (11) | 60 | C-7A | 6 15/16 | 5 3/16 | 1000 | ... |
| | ▲Medium | 150/99CL | 230-250 | 1.08 | Clear Extended Service (11) | 60 | C-7A | 6 15/16 | 5 3/16 | 2500 | ... |
| PS-30 | | 150PS30 | 230-250 | 1.56 | Clear—Silvered Neck (11) | 60 | C-7A | 8 1/16 | 6 | 1000 | ... |
| **150-250-400 WATTS** | | | | | | | | | | | |
| PS-35 | 3-Contact Mogul | 150/400CL | 120 | 3.94 | Clear—Medical Spotlight. 3-Way (11) | 24 | 2C-7A | 9⅜ | 7 | 200 | ... |
| | | 150/400 | 120 | 4.05 | Inside Frosted Medical Spotlight. 3-Way (11) | 24 | 2C-7A | 9⅜ | 7 | 200 | ... |
| **200 WATTS** | | | | | | | | | | | |
| A-23  | Medium | 200A | 120 | .54 | Inside Frosted Bonus Line (46) | 60 | CC-8 | 6 3/16 | 4⅝ | 750 | 4010 |
| | | 200A 12PK PM | 120 | .54 | Inside Frosted—Bonus Line. 12-Pack **PRICE MARKED** (46) | 60 | CC-8 | 6 3/16 | 4⅝ | 750 | 4010 |
| | | 200A | 125-130 | .64 | Inside Frosted—Bonus Line (46) | 60 | CC-8 | 6 3/16 | 4⅝ | 750 | 4010 |
| | | 200A/W 12PK PM | 120 | .59 | Soft-White—Bonus Line 12-Pack **PRICE MARKED** (46) | 60 | CC-8 | 6 3/16 | ... | 750 | 3910 |
| | | 200A/CL | 120 | .54 | Clear—Bonus Line (46) | 60 | CC-8 | 6 3/16 | 4⅝ | 750 | 4010 |
| | | 200A/CL | 125-130 | .64 | Clear—Bonus Line (46) | 60 | CC-8 | 6 3/16 | 4⅝ | 750 | 4010 |
| | ▲Medium | 200A/99 | 120 | .82 | Inside Frosted—Extended Service (46) | 60 | CC-8 | 6 3/16 | 4⅝ | 2500 | 3410 |
| | | 200A/99 24PK | 120 | .82 | Inside Frosted—Extended Service. 24-Pack (46) | 120 | CC-8 | 6 3/16 | 4⅝ | 2500 | 3410 |

| | | | | | | | | | | | |
|---|---|---|---|---|---|---|---|---|---|---|---|
| | | **200A/99** | 125-130 | .98 | Inside Frosted—Extended Service (46) | 60 | CC-8 | 6 3/16 | 4 5/8 | 2500 | 3410 |
| | | **200A/99CL** | 120 | .82 | Clear—Extended Service (46) | 60 | CC-8 | 6 3/16 | 4 5/8 | 2500 | . . . |
| | | **200A/99CL** | 125-130 | .98 | Clear—Extended Service (46) | 60 | CC-8 | 6 3/16 | 4 5/8 | 2500 | . . . |
| PS-30  | Medium | **200** | 120 | .71 | Clear (11) | 60 | C-9 | 8 1/16 | 6 | 750 | 3710 |
| | | **200** | 125 | .77 | Clear (11) | 60 | C-9 | 8 1/16 | 6 | 750 | 3710 |
| | | **200** | 130 | .85 | Clear (11) | 60 | C-9 | 8 1/16 | 6 | 750 | 3710 |
| | | **200/IF** | 120 | .71 | Inside Frosted (11) | 60 | C-9 | 8 1/16 | 6 | 750 | 3710 |
| | ▲Medium | **200PS30/29TF** | 115-125 | 1.78 | Inside Frosted—TUFF-SKIN (44) | 60 | C-9 | 8 1/16 | . . . | 1000 | . . . |
| | Medium | **200/IF** | 125 | .77 | Inside Frosted (11) | 60 | C-9 | 8 1/16 | 6 | 750 | 3710 |
| | | **200/IF** | 130 | .85 | Inside Frosted (11) | 60 | C-9 | 8 1/16 | 6 | 750 | 3710 |
| | Mogul | **200PS30/12** | 120 | .95 | Clear (11) | 60 | C-9 | 8 7/16 | 6 3/8 | 750 | . . . |
| | | **200PS30/12** | 130 | 1.13 | Clear (11) | 60 | C-9 | 8 7/16 | 6 3/8 | 750 | . . . |
| | ▲Medium | **200/99** | 120 | .95 | Clear—Extended Service (11) | 60 | C-9 | 8 1/16 | 6 | 2500 | 3260 |
| | | **200/99 24PK** | 120 | .95 | Clear—Extended Service. 24-Pack (11) | 72 | C-9 | 8 1/16 | 6 | 2500 | 3260 |
| | | **200/99** | 125-130 | 1.13 | Clear—Extended Service (11) | 60 | C-9 | 8 1/16 | 6 | 2500 | 3260 |
| | | **200/99IF** | 120 | .95 | Inside Frosted—Extended Service (11) | 60 | C-9 | 8 1/16 | 6 | 2500 | . . . |
| | | **200/99IF 24PK** | 120 | .95 | Inside Frosted—Extended Service. 24-Pack (11) | 72 | C-9 | 8 1/16 | 6 | 2500 | . . . |
| | | **200/99IF** | 125-130 | 1.13 | Inside Frosted—Extended Service (11) | 60 | C-9 | 8 1/16 | 6 | 2500 | . . . |
| | Medium | **200PS30/23** | 120 | 1.29 | Inside Frosted—Rough Service (11) | 24 | C-9 | 8 1/16 | 6 | 1000 | 3400 |
| | ▲Medium | **200PS30/23TF** | 115-125 | 2.59 | Inside Frosted—Rough Service. TUFF-SKIN (44) | 60 | C-9 | 8 1/16 | . . . | 1000 | . . . |
| | Medium | **200PS30/23** | 125-130 | 1.55 | Inside Frosted—Rough Service (11) | 24 | C-9 | 8 1/16 | 6 | 1000 | 3400 |

## INCANDESCENT LAMPS

| Bulb | Base | Lamp Ordering Code | Volts | Approx. List Price | DESCRIPTION | Std. Pkg. Qty. | Filament Design | M. O. L. | L. C. L. | Approx. Hours Life | Approx. Initial Lumens |
|---|---|---|---|---|---|---|---|---|---|---|---|
| **200 WATTS** (continued) | | | | | | | | | | | |
| PS-30  | Medium | **200PS30/24** | 120 | **$1.18** | Clear—Rough Service (11) | 60 | C-9 | 8 1/16 | 6 | 1000 | 3400 |
| | | **200PS30/24** | 125-130 | **1.42** | Clear—Rough Service (11) | 60 | C-9 | 8 1/16 | 6 | 1000 | 3400 |
| | | **200/D** | 115-125 | **1.66** | Clear—Daylight (11) | 60 | C-9 | 8 1/16 | 6 | 1000 | . . . |
| | | **200/SBIF** | 120 | **1.45** | Inside Frosted—Silvered Bowl (3,14) | 60 | C-9 | 8 1/16 | 6 | 1000 | 3320 |
| | | **200/SBIF** | 125-130 | **1.74** | Inside Frosted—Silvered Bowl (3,14) | 60 | C-9 | 8 1/16 | 6 | 1000 | 3320 |
| ★PAR-46  | Medium Side Prong | **200PAR46/3NSP** | 120 | **6.95** | PAR—Narrow Spotlight (15,58) | 12 | CC-13 | 4 | . . . | 2000 | 2300 |
| | | **200PAR46/3NSP** | 125-130 | **8.35** | PAR—Narrow Spotlight (15,58) | 12 | CC-13 | 4 | . . . | 2000 | 2300 |
| | | **200PAR46/3MFL** | 120 | **6.95** | PAR—Medium Floodlight (15,58) | 12 | CC-13 | 4 | . . . | 2000 | 2300 |
| | | **200PAR46/3MFL** | 125-130 | **8.35** | PAR—Medium Floodlight (15,58) | 12 | CC-13 | 4 | . . . | 2000 | 2300 |
| R-40 | ▲Medium | **200R/FL** | 120 | **2.32** | Reflector Floodlight—I.F. (14,34,35) | 24 | C-11 | 6 9/16 | . . . | 2000 | . . . |
| | | **200R/FL** | 125-130 | **2.78** | Reflector Floodlight—I.F. (14,34,35) | 24 | C-11 | 6 9/16 | . . . | 2000 | . . . |
| | Medium Skirted | **200R/4FL** | 120 | **2.43** | Industrial Reflector Floodlight — I.F. (14,34) | 24 | C-11 | 7 3/4 | . . . | 2500 | 2240 |
| ★PAR-56 | Screw Terminal | **200PAR** | 12 | **8.15** | PAR—Locomotive Headlight (2) | 12 | CC-8 | 4 1/2 | . . . | 500 | . . . |
| | | **200PAR** | 30 | **6.95** | PAR—Locomotive Headlight (2) | 12 | CC-8 | 4 1/2 | . . . | 500 | . . . |
| PS-30  | Medium | **200** | 32 | **2.05** | Clear—Train (11) | 60 | C-9 | 8 1/16 | 6 | 1000 | . . . |
| | | **200/IF** | 230,250 | **1.00** | Inside Frosted (11) | 60 | C-9 | 8 1/16 | 6 | 1000 | 3100 |
| | | **200** | 230,250 | **1.00** | Clear (11) | 60 | C-9 | 8 1/16 | 6 | 1000 | 3100 |
| | | **200** | 277,300 | **1.24** | Clear (11) | 60 | C-9 | 8 1/16 | 6 | 1000 | . . . |

| | | | | | | | | | | |
|---|---|---|---|---|---|---|---|---|---|---|
| | ▲Medium | **200/99** | **230-250** | **1.24** | Clear—Extended Service (11) | 60 | C-7A | 8 1/16 | 6 | 2500 | ... |
| | | **200/99IF** | **230-250** | **1.24** | Inside Frosted—Extended Service (11) | 60 | C-7A | 8 1/16 | 6 | 2500 | ... |
| PS-30 | Medium | **200PS30/RS** | **250** | **1.56** | Clear—Rough Service (11) | 60 | C-9 | 8 1/16 | 6 | 1000 | ... |

## 240 WATTS

| | | | | | | | | | | |
|---|---|---|---|---|---|---|---|---|---|---|
| ★PAR-56 | Screw Terminal | **240PAR56VNSP** | **12** | **5.60** | PAR—Very Narrow Spotlight (15) | 12 | C-6 | 4 1/2 | ... | 2000 | ... |
| | | **240PAR56MFL** | **12** | **5.60** | PAR—Medium Floodlight (15) | 12 | C-6 | 4 1/2 | ... | 2000 | ... |
| | | **240PAR56WFL** | **12** | **5.60** | PAR—Wide Floodlight (15) | 12 | C-6 | 4 1/2 | ... | 2000 | ... |

## 250 WATTS

| | | | | | | | | | | |
|---|---|---|---|---|---|---|---|---|---|---|
| A-21 | Medium | **250A21/60** | **115-125** | **7.00** | Purple X-Intermittent burning only (16,27) | 120 | C-9 | 4 7/8 | 3 7/16 | 50 | ... |
| G-30 | Medium Skirted | **250G30** | **115-125** | **2.05** | Clear—Infrared—Industrial (6,14,16) | 60 | C-7A | 7 3/16 | 5 | 5000+ | ... |
| | ▲Medium | **250G/SP** | **120** | ◆ | Clear—Spotlight (57) | 60 | C-5 | 5 1/8 | 3 | 200 | 4500 |
| | | **250G/FL** | **120** | **3.51** | Clear—Floodlight (13) | 60 | C-5 | 5 1/8 | 3 | 800 | 3650 |
| P-25 | | **250P25** | **120** | **3.45** | Clear — Locomotive Headlight (13) | 60 | C-5 | 4 3/4 | 3 | 500 | 3700 |
| PS-30 | Medium | **250PS30/33** | **115-125** | **1.08** | Reflector Infrared. Brooder—Silvered Neck (6,11,16) | 60 | C-9 | 8 1/16 | 6 | 5000+ | ... |
| R-40  | | **250R40/1 PM** | **115-125** | **1.89** | Reflector-Infrared Heat **PRICE MARKED** (6,34,37) | 24 | C-9 | 6 9/16 | ... | 5000+ | ... |
| | | **250R40/1 6PK PM** | **115-125** | **1.89** | Reflector—Infrared Heat **PRICE MARKED** (6,34,37) | 30 | C-9 | 6 9/16 | ... | 5000+ | ... |
| ★R-40 | | **250R40/10 6PK PM** | **115-125** | **3.80** | Reflector—Infrared Heat —Red bowl **PRICE MARKED** (6,37) | 30 | C-9 | 6 3/4 | ... | 5000+ | ... |
| R-40 | Medium Skirted | **250R40/4** | **115-125** | **2.64** | Reflector Infrared—Industrial (6,14,34) | 24 | C-9 | 7 3/8 | ... | 5000+ | ... |
| ★R-40 | | **250R40/5** | **115-125** | **3.67** | Reflector Infrared—Industrial (6,14) | 24 | C-9 | 7 1/2 | ... | 5000+ | ... |

◆ Now classified as Stage/Studio Lamp type. For information on the complete line of Stage/Studio Lamps, consult your General Electric Stage/Studio Lamp supplier or the nearest GE Lamp Sales District Office or Distribution Center.

## INCANDESCENT LAMPS

| Bulb | Base | Lamp Ordering Code | Volts | Approx. List Price | DESCRIPTION | Std. Pkg. Qty. | Filament Design | M. O. L. | L. C. L. | Approx. Hours Life | Approx. Initial Lumens |
|---|---|---|---|---|---|---|---|---|---|---|---|
| **250 WATTS** | | | | | | | | | | | |
| ★T-12 | Medium Prefocus | 250T12/8 | 120 | ◆ | Clear—Spotlight (52) | 24 | C-13 | 6⅛ | 3½ | 800 | . . . |
| T-14 | ▲Medium | 250T14/1 | 120 | $6.25 | Clear—Lighthouse (1) | 24 | C-5 | 5½ | 3 | 800 | . . . |
| ★T-20 | Medium Prefocus | 250T20/47 | 120 | ◆ | Clear—Spotlight. Burn base down to horizontal | 24 | C-13 | 5¾ | 2 3/16 | 200 | . . . |
| P-25 | ▲Medium | 250P25 | 32 | 3.45 | Clear—Locomotive Headlight (13) | 60 | C-5A | 4¾ | 3 | 500 | 4650 |
| | Medium Prefocus | 250P25/22 | 32 | 4.64 | Clear—Locomotive Headlight (13) | 60 | C-5A | 5 | 2 3/16 | 500 | . . . |
| **300 WATTS** | | | | | | | | | | | |
| PS-25 | Medium | 300M | 120 | .83 | Clear—Bonus Line (11) | 60 | CC-8 | 6 15/16 | 5 3/16 | 750 | 6360 |
| | | 300M | 125-130 | .99 | Clear—Bonus Line (11) | 60 | CC-8 | 6 15/16 | 5 3/16 | 750 | 6360 |
| | | 300M/IF | 120 | .83 | Inside Frosted—Bonus Line (11) | 60 | CC-8 | 6 15/16 | 5 3/16 | 750 | 6360 |
| | | 300M/IF | 125-130 | .99 | Inside Frosted—Bonus Line (11) | 60 | CC-8 | 6 15/16 | 5 3/16 | 750 | 6360 |
| PS-30  | ▲Medium | 300M/99 | 120 | 1.05 | Clear—Extended Service (11) | 60 | C-9 | 8 1/16 | 6 | 2500 | 5190 |
| | | 300M/99  24PK | 120 | 1.05 | Clear—Extended Service 24-Pack (11) | 72 | C-9 | 8 1/16 | 6 | 2500 | 5190 |
| | | 300M/99 | 125-130 | 1.27 | Clear—Extended Service (11) | 60 | C-9 | 8 1/16 | 6 | 2500 | 5190 |
| | | 300M/99IF | 120 | 1.05 | Inside Frosted—Extended Service (11) | 60 | C-9 | 8 1/16 | 6 | 2500 | . . . |
| | | 300M/99IF 24PK | 120 | 1.05 | Inside Frosted—Extended Service 24-Pack (11) | 72 | C-9 | 8 1/16 | 6 | 2500 | . . . |
| | | 300M/99IF | 125-130 | 1.27 | Inside Frosted—Extended Service (11) | 60 | C-9 | 8 1/16 | 6 | 2500 | . . . |
| | Medium | 300M/PS30 | 120 | .88 | Clear (11) | 60 | C-9 | 8 1/16 | 6 | 750 | 6110 |
| | | 300M/PS30/IF | 120 | .88 | Inside Frosted (11) | 60 | C-9 | 8 1/16 | 6 | 750 | 6110 |
| | | 300M/PS30/IF | 125 | .97 | Inside Frosted (11) | 60 | C-9 | 8 1/16 | 6 | 750 | 6110 |
| | | 300M/PS30/IF | 130 | 1.05 | Inside Frosted (11) | 60 | C-9 | 8 1/16 | 6 | 750 | 6110 |

| | | | | | | | | | | | |
|---|---|---|---|---|---|---|---|---|---|---|---|
| PS-35  | Medium Skirted | **300MS/SBIF** | **120** | **2.70** | Inside Frosted—Silvered Bowl (3,14) | 24 | C-9 | 9⅞ | 7½ | 1000 | . . . |
| | Mogul | **300** | **120** | **1.05** | Clear (11) | 24 | C-9 | 9⅜ | 7 | 1000 | 5820 |
| | | **300** | **125** | **1.16** | Clear (11) | 24 | C-9 | 9⅜ | 7 | 1000 | 5820 |
| | | **300** | **130** | **1.27** | Clear (11) | 24 | C-9 | 9⅜ | 7 | 1000 | 5820 |
| | | **300/IF** | **120** | **1.05** | Inside Frosted (11) | 24 | C-9 | 9⅜ | 7 | 1000 | 5820 |
| | | **300/IF** | **125** | **1.16** | Inside Frosted (11) | 24 | C-9 | 9⅜ | 7 | 1000 | 5820 |
| | | **300/IF** | **130** | **1.27** | Inside Frosted (11) | 24 | C-9 | 9⅜ | 7 | 1000 | 5820 |
| | ▲Mogul | **300/99** | **120** | **1.24** | Clear—Extended Service (11) | 24 | C-9 | 9⅜ | 7 | 2500 | 5190 |
| | | **300/99** | **125-130** | **1.49** | Clear—Extended Service (11) | 24 | C-9 | 9⅜ | 7 | 2500 | 5190 |
| | | **300/99IF** | **120** | **1.24** | Inside Frosted—Extended Service (11) | 24 | C-9 | 9⅜ | 7 | 2500 | . . . |
| | | **300/99IF** | **125-130** | **1.49** | Inside Frosted—Extended Service (11) | 24 | C-9 | 9⅜ | 7 | 2500 | . . . |
| | | **300/RS** | **120** | **2.16** | Clear—Rough Service (11) | 24 | C-9 | 9⅜ | 7 | 1000 | 5340 |
| | | **300/RS** | **125-130** | **2.59** | Clear—Rough Service (11) | 24 | C-9 | 9⅜ | 7 | 1000 | 5340 |
| | Mogul | **300/SBIF** | **120** | **2.10** | Inside Frosted—Silvered Bowl (3,14) | 24 | C-9 | 9⅜ | 7 | 1000 | 5410 |
| | | **300/SBIF** | **125-130** | **2.52** | Inside Frosted—Silvered Bowl (3,14) | 24 | C-9 | 9⅜ | 7 | 1000 | 5410 |
| | | **300/SBIF/1** | **120** | **2.37** | Inside Frosted—Semi-Silvered Bowl (3,14) | 24 | C-9 | 9⅜ | 7 | 1000 | . . . |
| | | **300/SBIF/1** | **125-130** | **2.85** | Inside Frosted—Semi-Silvered Bowl (3,14) | 24 | C-9 | 9⅜ | 7 | 1000 | . . . |
| ★PAR-56  | Mogul End Prong | **300PAR56/NSP** | **120** | **7.35** | PAR—Narrow Spotlight (15,59) | 12 | CC-13 | 5 | . . . | 2000 | 3840 |
| | | **300PAR56/NSP** | **125-130** | **8.85** | PAR—Narrow Spotlight (15,59) | 12 | CC-13 | 5 | . . . | 2000 | 3840 |
| | | **300PAR56/MFL** | **120** | **7.35** | PAR—Medium Floodlight (15,59) | 12 | CC-13 | 5 | . . . | 2000 | 3840 |
| | | **300PAR56/MFL** | **125-130** | **8.85** | PAR—Medium Floodlight (15,59) | 12 | CC-13 | 5 | . . . | 2000 | 3840 |
| | | **300PAR56/WFL** | **120** | **7.35** | PAR—Wide Floodlight (15,59) | 12 | CC-13 | 5 | . . . | 2000 | 3840 |
| | | **300PAR56/WFL** | **125-130** | **8.85** | PAR—Wide Floodlight (15,59) | 12 | CC-13 | 5 | . . . | 2000 | 3840 |

◆ Now classified as Stage/Studio Lamp type. For information on the complete line of Stage/Studio Lamps, consult your General Electric Stage/Studio Lamp supplier or the nearest GE Lamp Sales District Office or Distribution Center.

## 300 WATTS (Continued)

| Bulb | Base | Lamp Ordering Code | Volts | Approx. List Price | DESCRIPTION | Std. Pkg. Qty. | Filament Design | M.O.L. | L.C.L. | Approx. Hours Life | Approx. Initial Lumens |
|---|---|---|---|---|---|---|---|---|---|---|---|
| ★PAR-56 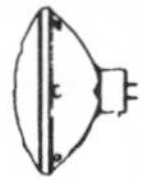 | Extended Mogul End Prong | 300PAR56/2NSP | 120 | $13.90 | PAR—Cool-Beam—Narrow Spotlight (18,59,77) | 12 | CC-13 | 5¾ | ... | 2000 | ... |
| | | 300PAR56/2NSP | 125-130 | 16.70 | PAR—Cool-Beam—Narrow Spotlight (18,59,77) | 12 | CC-13 | 5¾ | ... | 2000 | ... |
| | | 300PAR56/2MFL | 120 | 13.90 | PAR—Cool-Beam—Medium Floodlight (18,59,77) | 12 | CC-13 | 5¾ | ... | 2000 | ... |
| | | 300PAR56/2MFL | 125-130 | 16.70 | PAR—Cool-Beam—Medium Floodlight (18,59,77) | 12 | CC-13 | 5¾ | ... | 2000 | ... |
| | | 300PAR56/2WFL | 120 | 13.90 | PAR—Cool-Beam—Wide Floodlight (18,59,77) | 12 | CC-13 | 5¾ | ... | 2000 | ... |
| | | 300PAR56/2WFL | 125-130 | 16.70 | PAR—Cool-Beam—Wide Floodlight (18,59,77) | 12 | CC-13 | 5¾ | ... | 2000 | ... |
| R-40  | ▲Medium | 300R/FL | 120 | 2.48 | Reflector Floodlight—I.F. (14,34,35,36,73) | 24 | CC-2V Horiz. | 6⁹⁄₁₆ | ... | 2000 | 3650 |
| | | 300R/FL | 125-130 | 2.98 | Reflector Floodlight—I.F. (14,34,35,36,73) | 24 | CC-2V Horiz. | 6⁹⁄₁₆ | ... | 2000 | 3650 |
| ★R-40 | | 300R/FL/1 | 120 | 4.42 | Reflector Floodlight—I.F. (14,33,35,73) | 24 | CC-2V Horiz. | 6¾ | ... | 2000 | 3650 |
| | | 300R/FL/1 | 125-130 | 5.30 | Reflector Floodlight—I.F. (14,33,35,73) | 24 | CC-2V Horiz. | 6¾ | ... | 2000 | 3650 |
| | ▲Mogul | 300R/3FL | 120 | 4.80 | Reflector Floodlight—I.F. (33,73) | 24 | CC-2V Horiz. | 7¼ | ... | 2000 | 3650 |
| | | 300R/3FL | 125-130 | 5.75 | Reflector Floodlight—I.F. (33,73) | 24 | CC-2V Horiz. | 7¼ | ... | 2000 | 3650 |
| | | 300R/3FL/MS | 120 | 6.75 | Floodlight—I.F. Mill Service | 24 | C-7A | 7¼ | ... | 1000 | ... |
| R-40 | ▲Medium | 300R/SP | 120 | 2.48 | Reflector Spotlight—Light I.F. (14,34,35,36,73) | 24 | CC-2V Horiz. | 6⁹⁄₁₆ | ... | 2000 | 3650 |
| | | 300R/SP | 125-130 | 2.98 | Reflector Spotlight—Light I.F. (14,34,35,36,73) | 24 | CC-2V Horiz. | 6⁹⁄₁₆ | ... | 2000 | 3650 |

| | | | | | | | | | | | |
|---|---|---|---|---|---|---|---|---|---|---|---|
| ★R-40 | ▲Mogul | **300R/3SP** | 120 | **4.80** | Reflector Spotlight—Light I.F. (33,73) | 24 | CC-2V Horiz. | 7¼ | . . . | 2000 | 3650 |
| ★T-20 | Medium Bipost | **300T20/1** | 120 | **11.30** | Inside Frosted. Burn base up (73) | 12 | C-13 | 6½ | 4 | 1000 | . . . |
| | | **300T20/1** | 125-130 | **13.60** | Inside Frosted. Burn base up (73) | 12 | C-13 | 6½ | 4 | 1000 | . . . |
| P-25 | Medium Prefocus | **300P25P** | 60 | **6.40** | Clear—Locomotive Headlight (57) | 60 | C-5 | 5 | 2 3/16 | 500 | . . . |
| PS-30 | Medium | **300M/1** | 250 | **1.29** | Clear (11) | 60 | C-7A | 8 1/16 | 6 | 1000 | . . . |
| PS-35 | Medium Skirted | **300MS** | 250 | **2.16** | Clear (11) | 24 | C-7A | 9 7/8 | 7½ | 1000 | . . . |
| | ▲Mogul | **300** | 250 | **1.67** | Clear (11) | 24 | C-7A | 9 3/8 | 7 | 1000 | 4890 |
| | | **300/IF** | 250 | **1.67** | Inside Frosted (11) | 24 | C-7A | 9 3/8 | 7 | 1000 | 4890 |
| | | **300/99** | 230-250 | **1.94** | Clear—Extended Service (11) | 24 | C-7A | 9 3/8 | 7 | 2500 | . . . |
| | | **300** | 277 | **2.16** | Clear (11) | 24 | C-7A | 9 3/8 | 7 | 1000 | . . . |
| ★R-40 | | **300R/3FL** | 230,250 | **7.10** | Reflector Floodlight—I.F. (33,73) | 24 | C-7A | 7¼ | . . . | 2000 | . . . |
| | | **300R/3FL** | 277 | **7.10** | Reflector Floodlight—I.F. (33,73) | 24 | C-7A | 7¼ | . . . | 2000 | . . . |

## 375 WATTS

| | | | | | | | | | | | |
|---|---|---|---|---|---|---|---|---|---|---|---|
| G-30 | Medium Skirted | **375G30** | 115-125 | **2.48** | Clear—Infrared—Industrial (6,14,16) | 24 | C-7A | 7 3/16 | 5 | 5000+ | . . . |
| R-40 | | **375R40** | 115-125 | **3.24** | Reflector Infrared—Industrial. Light I. F. (6,14,16,34) | 24 | C-9 | 7 3/8 | . . . | 5000+ | . . . |
| ★R-40 | | **375R40/1** | 115-125 | **4.40** | Reflector Infrared — Industrial — Clear (6,14,16) | 24 | C-9 | 7½ | . . . | 5000+ | . . . |
| | | **375R40/10** | 115-125 | **5.80** | Reflector Infrared — Industrial — Red Bowl (6,14,16) | 24 | C-9 | 7½ | . . . | 5000+ | . . . |

## 400 WATTS

| | | | | | | | | | | | |
|---|---|---|---|---|---|---|---|---|---|---|---|
| G-30 | ▲Medium | **400G/SP** | 120 | ◆ | Clear—Spotlight (57) | 60 | C-5 | 5 1/8 | 3 | 200 | 8400 |
| | | **400G/FL** | 120 | **3.75** | Clear—Floodlight (57) | 60 | C-5 | 5 1/8 | 3 | 800 | 6800 |
| | | **400G/FL** | 130 | **4.50** | Clear—Floodlight (57) | 60 | C-5 | 5 1/8 | 3 | 800 | 6800 |

◆ Now classified as Stage/Studio Lamp type. For information on the complete line of Stage/Studio Lamps, consult your General Electric Stage/Studio Lamp supplier or the nearest GE Lamp Sales District Office or Distribution Center.

# INCANDESCENT LAMPS

| Bulb | Base | Lamp Ordering Code | Volts | Approx. List Price | DESCRIPTION | Std. Pkg. Qty. | Filament Design | M. O. L. | L. C. L. | Approx. Hours Life | Approx. Initial Lumens |
|---|---|---|---|---|---|---|---|---|---|---|---|
| **500 WATTS** | | | | | | | | | | | |
| PS-35  | ▲Mogul | **500** | 120 | $1.45 | Clear—Bonus Line (46) | 24 | CC-8 | 9⅜ | 7 | 1000 | 10850 |
| | | **500** | 125 | 1.59 | Clear—Bonus Line (46) | 24 | CC-8 | 9⅜ | 7 | 1000 | 10850 |
| | | **500** | 130 | 1.74 | Clear (46) | 24 | CC-8 | 9⅜ | 7 | 1000 | 10850 |
| | | **500/IF** | 120 | 1.51 | Inside Frosted—Bonus Line (46) | 24 | CC-8 | 9⅜ | 7 | 1000 | 10850 |
| | | **500/IF** | 130 | 1.81 | Inside Frosted—Bonus Line (46) | 24 | CC-8 | 9⅜ | 7 | 1000 | 10850 |
| PS-40  | | **500PS40** | 120 | 1.56 | Clear (11) | 24 | C-9 | 9¾ | 7 | 1000 | 9900 |
| | | **500PS40** | 125 | 1.72 | Clear (11) | 24 | C-9 | 9¾ | 7 | 1000 | 9900 |
| | | **500PS40** | 130 | 1.87 | Clear (11) | 24 | C-9 | 9¾ | 7 | 1000 | 9900 |
| | | **500PS40/IF** | 120 | 1.62 | Inside Frosted (11) | 24 | C-9 | 9¾ | 7 | 1000 | 9900 |
| | | **500PS40/IF** | 125 | 1.78 | Inside Frosted (11) | 24 | C-9 | 9¾ | 7 | 1000 | 9900 |
| | | **500PS40/IF** | 130 | 1.94 | Inside Frosted (11) | 24 | C-9 | 9¾ | 7 | 1000 | 9900 |
| | | **500/99** | 120 | 1.83 | Clear—Extended Service (11) | 24 | C-9 | 9¾ | 7 | 2500 | 9070 |
| | | **500/99** | 130 | 2.20 | Clear—Extended Service (11) | 24 | C-9 | 9¾ | 7 | 2500 | 9070 |
| | | **500/99IF** | 120 | 1.94 | Inside Frosted—Extended Service (11) | 24 | C-9 | 9¾ | 7 | 2500 | 9070 |
| | | **500/99IF** | 130 | 2.33 | Inside Frosted—Extended Service (11) | 24 | C-9 | 9¾ | 7 | 2500 | 9070 |
| | | **500/RS** | 120 | 2.26 | Clear—Rough Service (11) | 24 | C-9 | 9¾ | 7 | 1000 | ... |
| | | **500/RS** | 125-130 | 2.72 | Clear—Rough Service (11) | 24 | C-9 | 9¾ | 7 | 1000 | ... |
| | | **500/SBIF** | 120 | 2.86 | Inside Frosted—Silvered Bowl. Burn base up (14) | 24 | C-9 | 9¾ | 7 | 1000 | 9530 |
| | | **500/SBIF** | 130 | 3.43 | Inside Frosted—Silvered Bowl. Burn base up (14) | 24 | C-9 | 9¾ | 7 | 1000 | 9530 |
| G-30 | Medium Skirted | **500G30/1** | 115-125 | 2.53 | Clear—Infrared. Industrial (6,14,16) | 60 | C-7A | 7³⁄₁₆ | 5 | 5000+ | ... |

| | | | | | | | | | | | |
|---|---|---|---|---|---|---|---|---|---|---|---|
| G-40 | ▲Mogul | **500G/FL** | 120 | 5.80 | Clear—Floodlight (57) | 24 | C-5 | 7 5/16 | 4¼ | 800 | 9300 |
| ★PAR-64 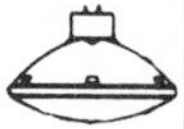 | Extended Mogul End Prong | **500PAR64/NSP** | 120 | 16.20 | PAR—Narrow Spotlight (18,59) | 12 | CC-13 | 6 | . . . | 2000 | . . . |
| | | **500PAR64/NSP** | 125-130 | 19.40 | PAR—Narrow Spotlight (18,59) | 12 | CC-13 | 6 | . . . | 2000 | . . . |
| | | **500PAR64/MFL** | 120 | 16.20 | PAR—Medium Floodlight (18,59) | 12 | CC-13 | 6 | . . . | 2000 | . . . |
| | | **500PAR64/MFL** | 125-130 | 19.40 | PAR—Medium Floodlight (18,59) | 12 | CC-13 | 6 | . . . | 2000 | . . . |
| | | **500PAR64/WFL** | 120 | 16.20 | PAR—Wide Floodlight (18,59) | 12 | CC-13 | 6 | . . . | 2000 | . . . |
| | | **500PAR64/WFL** | 125-130 | 19.40 | PAR—Wide Floodlight (18,59) | 12 | CC-13 | 6 | . . . | 2000 | . . . |
| ★R-40  | ▲Mogul | **500R/3FL** | 120 | 5.75 | Reflector Floodlight—I.F. Collector Grid (24,33) | 24 | CC-2V<br>Horiz. | 7¼ | . . . | 2000 | 6500 |
| | | **500R/3FL** | 125-130 | 6.90 | Reflector Floodlight—I.F. Collector Grid (24,33) | 24 | CC-2V<br>Horiz. | 7¼ | . . . | 2000 | 6500 |
| | | **500R/3FL/MS** | 120 | 6.80 | Reflector Floodlight—I.F. Mill Service (24) | 24 | C-7A | 7¼ | . . . | 1000 | . . . |
| | | **500R/3FL/MS** | 125-130 | 8.15 | Reflector Floodlight—I.F. Mill Service (24) | 24 | C-7A | 7¼ | . . . | 1000 | . . . |
| | | **500R/3SP** | 120 | 5.75 | Reflector Spotlight—Light I.F. Collector Grid (24,33) | 24 | CC-2V<br>Horiz. | 7¼ | . . . | 2000 | 6500 |
| R-52 | | **500R52** | 120 | 6.25 | Reflector High Bay—Light I.F. Collector Grid (19,22) | 12 | C-7A | 11¾ | . . . | 2000 | 7600 |
| | | **500R52** | 125-130 | 7.50 | Reflector High Bay—Light I.F. Collector Grid (19,22) | 12 | C-7A | 11¾ | . . . | 2000 | 7600 |
| ★T-20 | Medium Bipost | **500T20/13** | 120 | 14.85 | Clear—Airway Beacon (1) | 12 | C-13B | 7½ | 3 | 500 | 10300 |
| | | **500T20/13** | 130 | 17.80 | Clear—Airway Beacon (1) | 12 | C-13B | 7½ | 3 | 500 | 10300 |
| | | **500T20/50** | 120 | 11.60 | Inside Frosted. Burn base up (73) | 12 | C-13 | 6½ | 4 | 1000 | 9800 |
| | | **500T20/50** | 125-130 | 13.90 | Inside Frosted. Burn base up (73) | 12 | C-13 | 6½ | 4 | 1000 | 9800 |
| | Medium Prefocus | **500T20/49** | 120 | 16.30 | Clear—Marine Fire Detector. Burn base down | 24 | C-13D | 5¾ | 2 3/16 | 200 | . . . |
| ★T-12 | | **500T12/8** | 120 | ♦ | Clear—Spotlight (8,52) | 24 | C-13D | 6⅛ | 3½ | 800 | . . . |
| | | **500T12/9** | 120 | ♦ | Clear—Spotlight (8,52) | 24 | C-13D | 6⅛ | 3½ | 200 | 11000 |
| ★T-14 | Medium Bipost | **500T14/7** | 120 | ♦ | Clear—Spotlight (17,52) | 24 | C-13D | 6 7/16 | 4 | 800 | . . . |
| | | **500T14/8** | 120 | ♦ | Clear—Spotlight (17,52) | 24 | C-13D | 6 7/16 | 4 | 200 | . . . |

♦ Now classified as Stage/Studio Lamp type. For information on the complete line of Stage/Studio Lamps, consult your General Electric Stage/Studio Lamp supplier or the nearest GE Lamp Sales District Office or Distribution Center

## T-12 Approx. 1½" Diameter

To determine approximate lamp length (pin ends to pin ends), deduct 1/4" from the nominal lamp length shown below.

| Nominal Lamp Watts | Bulb | Nominal Length (Inches) | Base | Lamp Ordering Code | Approx. List Price | DESCRIPTION | Std. Pkg. Qty. | Approx. Hours Life | Approx. Initial Lumens | Approx. Lumens at 40% Rated Avg. Life |
|---|---|---|---|---|---|---|---|---|---|---|
| 20 | T-12 | 24 | Medium Bipin | F20T12/WWX 6PK PM | $2.25 | DeLuxe Warm White. Improved Color. 6-Pack PRICE MARKED | 24 | 9000 | 820 | 715 |
| | | | | F20T12/D | 1.50 | Daylight | 24 | 9000 | 1075 | 890 |
| | | | | F20T12/D 6PK | 1.60 | Daylight. 6-Pack | 24 | 9000 | 1075 | 890 |
| | | | | F20T12/W | 1.50 | White | 24 | 9000 | 1300 | 1155 |
| | | | | F20T12/WW | 1.50 | Warm White | 24 | 9000 | 1300 | 1155 |
| | | | | F20T12/C50 | 2.30 | Chroma 50—Chromaline™. 5000°K. | 24 | 9000 | 850 | 755 |
| | | | | F20T12/N 6PK | 2.35 | Natural (Formerly Soft White-Natural). 6-Pack | 24 | 9000 | 810 | 705 |
| | | | | F20T12/PL 6PK | 3.65 | Plant Light. 6-Pack | 24 | 9000 | 340 | ... |
| | | | | F20T12/B 6PK | 3.25 | Blue. 6-Pack | 24 | 9000 | 450 | 325 |
| | | | | F20T12/G 6PK | 3.25 | Green. 6-Pack | 24 | 9000 | 1850 | ... |
| | | | | F20T12/GO 6PK | 3.25 | Gold. 6-Pack | 24 | 9000 | 900 | 650 |
| | | | | F20T12/PK 6PK | 3.25 | Pink. 6-Pack | 24 | 9000 | 450 | 250 |
| | | | | F20T12/R 6PK | 3.25 | Red. 6-Pack | 24 | 9000 | 80 | 45 |
| | | | | F20T12/BL 6PK | 2.70 | Black Light. 6-Pack | 24 | 9000 | ... | ... |
| | | | | F20T12/BLB 6PK PM | 16.00 | Black Light Blue. Integral Filter. 6-Pack PRICE MARKED | 24 | 9000 | ... | ... |
| | | | | F20T12/CW/1 | 2.45 | Cool White—D.C. Operation | 24 | 7500 | 995 | 895 |
| | | 26 | | F20T12/CW/26 | 2.45 | Cool White—Appliance Service | 24 | 7500 (343) | 1250 | 1000 |
| 25 | | 28 | | F25T12/CW/28 | 2.80 | Cool White | 24 | 7500 | 1700 | 1450 |
| | | | | F25T12/CW/28 6PK | 2.90 | Cool White. 6-Pack | 24 | 7500 | 1700 | 1450 |

| Type | Price | Description | Qty | Life | | |
|---|---|---|---|---|---|---|
| F25T12/D/28 | 3.00 | Daylight | 24 | | 1400 | 1150 |
| F25T12/CW/33 | 2.95 | Cool White | 24 | | 1815 | 1540 |
| F25T12/CW/33 6PK | 3.05 | Cool White. 6-Pack | 24 | | 1815 | 1540 |
| F25T12/D/33 | 3.15 | Daylight | 24 | | 1600 | 1450 |
| F25T12/W/33 | 3.15 | White | 24 | | 1850 | 1570 |
| F25T12/WW/33 | 3.15 | Warm White | 24 | | 1850 | 1570 |

## T-8 Approx. 1″ Diameter
To determine approximate lamp length (pin ends to pin ends), deduct 1/4″ from the nominal lamp length shown below.

Medium Bipin

| Type | Price | Description | Qty | Life | | |
|---|---|---|---|---|---|---|
| F30T8/CW | 1.80 | Cool White | 24 | 7500 | 2200 | 1875 |
| F30T8/CW 6PK | 1.95 | Cool White. 6-Pack | 24 | 7500 | 2200 | 1875 |
| F30T8/CWX | 2.65 | DeLuxe Cool White | 24 | 7500 | 1580 | 1300 |
| F30T8/CWX 6PK PM | 2.80 | DeLuxe Cool White. 6-Pack PRICE MARKED | 24 | 7500 | 1580 | 1300 |
| F30T8/WWX | 2.65 | DeLuxe Warm White | 24 | 7500 | 1520 | 1250 |
| F30T8/WWX 6PK PM | 2.80 | DeLuxe Warm White. 6-Pack PRICE MARKED | 24 | 7500 | 1520 | 1250 |
| F30T8/D | 2.00 | Daylight | 24 | 7500 | 1850 | 1570 |
| F30T8/D 6PK | 2.15 | Daylight. 6-Pack | 24 | 7500 | 1850 | 1570 |
| F30T8/W | 2.00 | White | 24 | 7500 | 2180 | 1850 |
| F30T8/WW | 2.00 | Warm White | 24 | 7500 | 2180 | 1850 |
| F30T8/N | 2.65 | Natural (Formerly Soft White-Natural) | 24 | 7500 | 1500 | 1200 |
| F30T8/N 6PK | 2.80 | Natural (Formerly Soft White-Natural). 6-Pack | 24 | 7500 | 1500 | 1200 |
| F30T8/PL 6PK | 3.70 | Plant Light. 6-Pack | 24 | 7500 | 630 | … |
| F30T8/B 6PK | 3.20 | Blue. 6-Pack | 24 | 7500 | 840 | 630 |
| F30T8/GO 6PK | 3.20 | Gold. 6-Pack | 24 | 7500 | 1470 | 950 |
| F30T8/PK 6PK | 3.20 | Pink. 6-Pack | 24 | 7500 | 840 | 460 |
| F30T8/BL 6PK | 2.85 | Black Light. 6-Pack | 24 | 7500 | … | … |
| F30T8/BLB 6PK | 16.40 | Black Light Blue. Integral Filter. 6-Pack | 24 | 7500 | … | … |

# PREHEAT FLUORESCENT LAMPS

| Nominal Lamp Watts | Bulb | Nominal Length (Inches) | Base | Lamp Ordering Code | Approx. List Price | DESCRIPTION | Std. Pkg. Qty. | Approx. Hours Life | Approx. Initial Lumens | Approx. Lumens at 40% Rated Avg.Life |
|---|---|---|---|---|---|---|---|---|---|---|

## T-17  Approx. 2¹⁄ₛ″ Diameter
To determine approximate lamp length (pin ends to pin ends), deduct 1/2″ from the nominal lamp length shown below.

| 90 | T-17 | 60 | Mogul Bipin | F90T17/CW | $3.45 | Cool White | 12 | 9000 § | 6400 | 5890 |
| | | | | F90T17/D | 3.65 | Daylight | 12 | 9000 § | 5650 | 5085 |
| | | | | F90T17/W | 3.65 | White | 12 | 9000 § | 6350 | 5840 |
| | | | | F90T17/WW | 3.65 | Warm White | 12 | 9000 § | 6350 | 5840 |

§ Rated average life at 3 hours per start. At 12 hours per start, rated average life is 13,500 hours.

## 30-WATT RAPID START LAMPS (No Starters Used)

### T-12  Approx. 1½″ Diameter
To determine approximate lamp length (pin ends to pin ends), deduct 1/4″ from the nominal lamp length shown below.

| 30 | T-12 | 36 | Medium Bipin | F30T12/CW/RS | $2.35 | Cool White | 24 | 15000 | 2300 | 2000 |
| | | | | F30T12/CW/RS  6PK | 2.42 | Cool White. 6-Pack | 24 | 15000 | 2300 | 2000 |
| | | | | F30T12/CWX/RS | 2.60 | DeLuxe Cool White | 24 | 15000 | 1530 | 1165 |
| | | | | F30T12/WWX/RS | 2.60 | DeLuxe Warm White. Improved Color | 24 | 15000 | 1480 | 1125 |
| | | | | F30T12/D/RS | 2.45 | Daylight | 24 | 15000 | 1950 | 1695 |
| | | | | F30T12/W/RS | 2.45 | White | 24 | 15000 | 2360 | 2055 |
| | | | | F30T12/WW/RS | 2.45 | Warm White | 24 | 15000 | 2360 | 2055 |
| | | | | F30T12/C50/RS | 2.80 | Chroma 50—Chromaline™— 5000°K. | 24 | 15000 | 1630 | ... |
| | | | | F30T12/N/RS | 2.85 | Natural (formerly Soft White — Natural) | 24 | 15000 | 1550 | ... |

## 40-WATT RAPID START—PREHEAT LAMPS

Since any one design cannot answer the full range of the economic requirements of the fluorescent lamp purchaser, General Electric offers both the **STAYBRIGHT**™ and **MAINLIGHTER**™ 40-Watt Rapid Start-Preheat Lamps.

The premium F40 STAYBRIGHT Lamp, with its "floating shield" electrode design, eliminates "end blackening" for better appearance, but more importantly, provides higher initial light output and better lumen maintenance and thus greater value, to be the most efficient 40-watt fluorescent lamp currently available.

The standard MAINLIGHTER Lamp offers the unmatched combination of more light, longer life at lowest operating costs to provide greater lighting value than any other standard 40-watt fluorescent lamp.

## STAYBRIGHT™ LAMPS

Life ratings published below for Staybright Lamps are for rapid start circuits only. On preheat circuits, rated average life at 3 hours per start is 12,000 hours; at 12 hours per start—14,000 hours.

### T-12    Approx. 1½" Diameter

**To determine approximate lamp length (pin ends to pin ends), deduct 1/4" from the nominal lamp length shown below.**

| Nominal Lamp Watts | Bulb | Nominal Length (Inches) | Base | | Lamp Ordering Code | | List Price | DESCRIPTION | Std. Pkg. Qty. | Approx. Hours Life | Approx. Initial Lumens | Approx. Lumens at 40% Rated Avg.Life |
|---|---|---|---|---|---|---|---|---|---|---|---|---|
| 40 | T-12 | 48 | Medium Bipin | | F40CW/S | | $1.49 | Cool White—STAYBRIGHT | 24 | 15000† | 3250 | 2960 |
| | | | | | F40CW/S | 6PK | 1.67 | Cool White—STAYBRIGHT 6-Pack | 24 | 15000† | 3250 | 2960 |
| | | | | | F40D/S | | 1.70 | Daylight—STAYBRIGHT | 24 | 15000† | 2650 | 2410 |
| | | | | | F40W/S | | 1.70 | White—STAYBRIGHT | 24 | 15000† | 3300 | 3005 |
| | | | | | F40WW/S | | 1.70 | Warm White—STAYBRIGHT | 24 | 15000† | 3300 | 3005 |

† Rated average life at 3 hours per start on rapid start circuits. 20,000 hours life at 12 hours per start.

# FLUORESCENT LAMPS
## 40-WATT RAPID START—PREHEAT LAMPS

### MAINLIGHTER™ LAMPS

Life ratings published below for Mainlighter Lamps are for rapid start circuits only. On preheat circuits, rated average life at 3 hours per start is 12,000 hours; at 12 hours per start—15,000 hours. All F40 Mainlighter Lamps are excellent alternates for F40T12/LT lamps.

## T-12 Approx. 1½″ Diameter

To determine approximate lamp length (pin ends to pin ends), deduct 1/4″ from the nominal lamp length shown below.

| Nominal Lamp Watts | Bulb | Nominal Length (Inches) | Base | Lamp Ordering Code | Approx. List Price | DESCRIPTION | Std. Pkg. Qty. | Approx. Hours Life | Approx. Initial Lumens | Approx. Lumens at 40% Rated Avg. Life |
|---|---|---|---|---|---|---|---|---|---|---|
| 40 | T-12 | 48 | Medium Bipin | F40CW | $1.29 | Cool White—MAINLIGHTER. | 24 | *18000+ | 3150 | 2805 |
| | | | | F40CW 6PK | 1.45 | Cool White—MAINLIGHTER. 6-Pack | 24 | *18000+ | 3150 | 2805 |
| | | | | F40CW 48PK | 1.29 | Cool White—MAINLIGHTER. 48-Pack | 48 | *18000+ | 3150 | 2805 |
| | | | | F40CW 540-PALLET | 1.29 | Cool White—MAINLIGHTER. Palletized—540 Lamps | 540 | *18000+ | 3150 | 2805 |
| | | | | F40CWX | 1.50 | DeLuxe Cool White—MAINLIGHTER | 24 | *18000+ | 2200 | 1850 |
| | | | | F40CWX 6PK PM | 1.65 | DeLuxe Cool White—MAINLIGHTER. 6-Pack PRICE MARKED | 24 | *18000+ | 2200 | 1850 |
| | | | | F40WWX | 1.50 | DeLuxe Warm White—Improved Color. MAINLIGHTER | 24 | *18000+ | 2150 | 1795 |
| | | | | F40WWX 6PK PM | 1.65 | DeLuxe Warm White—Improved Color. MAINLIGHTER. 6-Pack PRICE MARKED | 24 | *18000+ | 2150 | 1795 |
| | | | | F40D | 1.50 | Daylight—MAINLIGHTER | 24 | *18000+ | 2600 | 2315 |

| Code | Pack | Price | Description | | | | |
|---|---|---|---|---|---|---|---|
| F40D | 6PK | 1.65 | Daylight—MAINLIGHTER. 6-Pack | 24 | *18000+ | 2600 | 2315 |
| F40W | | 1.50 | White—MAINLIGHTER | 24 | *18000+ | 3200 | 2850 |
| F40WW | | 1.50 | Warm White—MAINLIGHTER | 24 | *18000+ | 3200 | 2850 |
| F40WW | 6PK | 1.65 | Warm White—MAINLIGHTER. 6-Pack | 24 | *18000+ | 3200 | 2850 |
| F40WW | 540-PALLET | 1.50 | Warm White—MAINLIGHTER. Palletized—540 Lamps | 540 | *18000+ | 3200 | 2850 |
| F40SGN | | 1.95 | Sign White—MAINLIGHTER | 24 | *18000+ | 2440 | 2050 |
| F40/C50 | | 1.95 | Chroma 50—Chromaline™—5000°K. MAINLIGHTER | 24 | *18000+ | 2150 | 1805 |
| F40/C75 | | 1.95 | Chroma 75—Chromaline™—7500°K. MAINLIGHTER | 24 | *18000+ | 1990 | 1630 |
| F40N | | 1.90 | Natural (formerly Soft White—Natural) MAINLIGHTER | 24 | *18000+ | 2100 | 1765 |
| F40N | 6PK | 2.00 | Natural (formerly Soft White—Natural) MAINLIGHTER. 6-Pack | 24 | *18000+ | 2100 | 1765 |
| F40CG | 6PK | 2.00 | Cool Green—MAINLIGHTER. 6-Pack | 24 | *18000+ | 2850 | 2395 |
| F40PL | 6PK | 4.60 | Plant Light—MAINLIGHTER. 6-Pack | 24 | *18000+ | 900 | 675 |
| F40B | 6PK | 3.10 | Blue—MAINLIGHTER. 6-Pack | 24 | *18000+ | 1160 | 740 |
| F40G | 6PK | 3.10 | Green—MAINLIGHTER. 6-Pack | 24 | *18000+ | 4500 | 2340 |
| F40GO | 6PK | 3.10 | Gold—MAINLIGHTER. 6-Pack | 24 | *18000+ | 2400 | 1800 |
| F40PK | 6PK | 3.10 | Pink—MAINLIGHTER. 6-Pack | 24 | *18000+ | 1160 | 440 |
| F40R | 6PK | 3.10 | Red—MAINLIGHTER. 6-Pack | 24 | *18000+ | 200 | 60 |
| F40BL | 6PK | 3.45 | Black Light. 6-Pack | 24 | *18000+ | ... | ... |
| F40BLB | 6PK PM | 18.25 | Black Light Blue—Integral Filter. 6-Pack. **PRICE MARKED** | 24 | *18000+ | ... | ... |
| FR40CW | | 3.70 | Cool White—Reflector. MAINLIGHTER. 135°window (308) | 24 | *18000+ | 2600 | ... |

* Rated average life at 3 hours per start. 25,000 + hours life at 12 hours per start; 30,000 + hours on continuous burning.

# FLUORESCENT LAMPS
## MOD-U-LINE™ U-SHAPED FLUORESCENT LAMPS (Rapid Start)

General Electric MOD-U-LINE "U-Shaped" Fluorescent Lamps are available in two leg spacings—3⅝" and 6". Approximate lamp length (base face to outside of glass bend) on F40/U and F40/U/3 lamps (with 3⅝" leg spacing) is 22¼ inches. On F40/U/6 lamps with 6" leg spacing, approximate lamp length (base face to outside of glass bend) is 22½ inches.

| Nominal Lamp Watts | Bulb | Nominal Length (Inches) | Base | Lamp Ordering Code | Approx. List Price | DESCRIPTION | Std. Pkg. Qty. | Approx. Hours Life | Approx. Initial Lumens | Approx. Lumens at 40% Rated Avg.Life |
|---|---|---|---|---|---|---|---|---|---|---|
| **T-12  Approx. 1½" Diameter** | | | | | | | | | | |
| 40 | T-12 U-shape | 24 | Medium Bipin | F40/CW/U/3 | $3.65 | Cool White—Mod-U-Line. 3⅝" Leg Spacing | 12 | 12000 ‡ | 2800 | 2520 |
| | | | | F40CW/U/3  576-PALLET | 3.65 | Cool White—Mod-U-Line. Palletized–576 Lamps (338) | 576 | 12000 ‡ | 2800 | 2520 |
| | | | | F40WW/U | 3.85 | Warm White—Mod-U-Line. 3⅝" Leg Spacing (338) | 6 | 12000 ‡ | 2800 | 2520 |
| | | | | F40WW/U/3 | 3.85 | Warm White—Mod-U-Line. 3⅝" Leg Spacing | 12 | 12000 ‡ | 2800 | 2520 |
| | | | | F40WW/U/3  576-PALLET | 3.85 | Warm White—Mod-U-Line. 3⅝" Leg Spacing. Palletized–576 Lamps (338) | 576 | 12000 ‡ | 2800 | 2520 |
| | | | | F40W/U/3 | 3.85 | White—Mod-U-Line. 3⅝" Leg Spacing | 12 | 12000 ‡ | 2800 | 2520 |
| | | | | F40CW/U/6 | 3.65 | Cool White—Mod-U-Line. 6" Leg Spacing | 12 | 12000 ‡ | 2800 | 2520 |
| | | | | F40WW/U/6 | 3.85 | Warm White—Mod-U-Line. 6" Leg Spacing | 12 | 12000 ‡ | 2800 | 2520 |
| | | | | F40W/U/6 | 3.85 | White—Mod-U-Line. 6" Leg Spacing | 12 | 12000 ‡ | 2800 | 2520 |

‡ Rated Average life at 3 hours per start. At 12 hours per start, rated average life is 18,000 hours.

## 40-WATT INSTANT START LAMPS (No Starters Used)

**NOTE:** The base pins on "40-Watt Instant Start" Fluorescent lamps are short-circuited inside the end caps and the lamps will not operate on preheat or rapid start circuits.

## T-12  Approx. 1½" Diameter
To determine approximate lamp length (pin ends to pin ends), deduct 1/4" from the nominal lamp length shown below.

| 40 | T-12 | 48 | Medium Bipin | F40T12/CW/IS | Cool White | $2.90 | 24 | 7500 | 3100 | 2850 |
|---|---|---|---|---|---|---|---|---|---|---|
| | | | | F40T12/CWX/IS | DeLuxe Cool White | 3.25 | 24 | 7500 | 2100 | 1760 |
| | | | | F40T12/D/IS | Daylight | 3.05 | 24 | 7500 | 2500 | 2250 |
| | | | | F40T12/W/IS | White | 3.05 | 24 | 7500 | 3100 | 2800 |
| | | | | F40T12/WW/IS | Warm White | 3.05 | 24 | 7500 | 3100 | 2800 |

## 17  Approx. 2⅛" Diameter
To determine approximate lamp length (pin ends to pin ends), deduct 1/2" from the nominal lamp length shown below.

| 40 | T-17 | 60 | Mogul Bipin | F40T17/CW/IS | Cool White | $5.35 | 12 | 7500 | 2950 | 2715 |
|---|---|---|---|---|---|---|---|---|---|---|
| | | | | F40T17/WW/IS | Warm White | 5.65 | 12 | 7500 | 2950 | 2715 |

## FLUORESCENT APPLIANCE LAMPS

Fluorescent Appliance Lamps are designed primarily for intermittent burning service on appliances. Life and watts depend upon the characteristics of the ballasts used within the individual circuits. Rated average life is 4500 hours at 45 minutes per start. **To determine approximate lamp length (pin ends to pin ends), deduct 1/4" from the nominal lamp length shown below.**

| Nominal Lamp Watts | Bulb | Nominal Length (Inches) | Base | Lamp Ordering Code | List Price | DESCRIPTION | Std. Pkg. Qty. | Estimated Average Life (Hours) | Approx. Initial Lumens | Approx. Lumens at 40% Rated Avg.Life |
|---|---|---|---|---|---|---|---|---|---|---|
| 18 | T-8 | 22 | Medium Bipin | F22"T8/D/4 | $2.95 | Daylight | 24 | 7500 | 900 | 735 |
| 19 | | 24 | | F24"T8/CW/4 | 3.10 | Cool White | 24 | 7500 | 1175 | 940 |
| | | 26 | | F26"T8/CW/4 | 3.10 | Cool White | 24 | 7500 | 1275 | 1020 |
| | | | | F26"T8/CW/4  6PK | 3.20 | Cool White. 6-Pack | 24 | 7500 | 1275 | 1020 |
| | | 28 | | F28"T8/CW/4 | 3.10 | Cool White | 24 | 7500 | 1350 | 1080 |
| | | 30 | | F30"T8/CW/4 | 3.10 | Cool White | 24 | 7500 | 1400 | 1120 |
| | | | | F30"T8/D/4 | 3.10 | Daylight | 24 | 7500 | 1300 | . . . |
| 21 | T-12 | 30 | | F30"T12/CW | 2.75 | Cool White | 24 | 7500 | 1400 | 1120 |
| 30 | T-8 | 34 | | F34"T8/CW/4 | 3.10 | Cool White | 24 | 7500 | 2000 | 1700 |

# Index

# Index

## A

Antenna head ................................................222
Antenna signal splitter ...............................226
Appliances, heating ......................................69
Appliances, motor operated .......................69
Appliances, portable ...................................160
Applications, practical .................................126
Automatic garage door system,
     installation .............................................300
Automatic garage door system,
     optional accessories ..........................304

## B

Boxes, outlet .................................................57

## C

Cable, armored ..............................................49
Cable, lead-in ...............................................224
Cable, nonmetallic sheath ...........................49
Cable, service-entrance ................................50
Cable systems ..............................................49
Cable, underground-feeder ..........................51
Chimes, location ..........................................246
Chime, two-note, installation ....................246
Chimes, types .............................................246

Chisels, wood ................................................................16
Circuit breakers ..........................................................203
Circuits, 240-volt, layout ...........................................120
Clothes dryer ..............................................................309
Clothes washer ...........................................................308
Compactor ..................................................................306
Condensing unit, mounting.........................................292
Cooling coil, installing................................................292
Couplers .....................................................................226

### D

Dishwasher .................................................................314
Disposal......................................................................307
Drill, electric ...............................................................16

### E

Electric barbecue........................................................315
Electric lamps ...............................................................68
Electric motor maintenance.........................................191
Electric motors,
     typical applications ..............................................196
Electric service, partial updating .................................149
Electric smoker ...........................................................315
Electrical faults, types...................................................61
Electrical service, correct size....................................133
Electrical wire joints, taping..........................................45
EMT, using ...................................................................54
Equipment grounding ..................................................158
Equipment grounding, summary ..................................162
Evaporative cooler.......................................................318

### F

Fault, partial ground ......................................................67
Firelogs.......................................................................317
Fittings, installing ........................................................266
Food center .................................................................316
Freezer .......................................................................308
Fuses............................................................................63
Fuses, cartridge...........................................................202
Fuses, plug..................................................................201

### H

Hacksaw blades ............................................................16
Hacksaw frame..............................................................16

Hammer, claw...................................................................15
Humidifier......................................................................317

## I

Intercom system,
    careful planning ........................................................233
Intercom system, installation ...........................................235

## L

Ladder, step....................................................................16
Light switches, installation...............................................90
Lighting fixtures, installation ...........................................171
Lighting fixtures, selecting .............................................168
Lightning arresters.........................................................214
Lightning rods ...............................................................208
Low-voltage remote-controlled
    switching................................................................112
Low-voltage switching,
    application ..............................................................114

## M

Master antenna TV system .............................................230
Masts ...........................................................................224
Microwave oven............................................................318
Motor, capacitor-start....................................................190
Motor, capacitor-start
    capacitor-run...........................................................190
Motor repairs ...............................................................198
Motor, repulsion-start
    induction-run............................................................191
Motor, shaded-pole.......................................................191
Motor, split-phase .........................................................190
Motor, universal............................................................191

## N

Neutral wire, loose..........................................................66

## O

Outdoor farm lighting, installing......................................176
Outlets, TV....................................................................228
Outside residential lighting,
    installing.................................................................184
Overcurrent protection,
    selecting proper.......................................................204

## P

Pliers, diagonal-cutting ..................................................14
Pliers, gripping ...........................................................21
Pliers, long-nose .........................................................13
Pliers, side-cutting ......................................................13
Pole metering .............................................................145
Preamplifier ...............................................................222

## R

Receptacles, calculating .............................................84
Refrigerant tubing, installing .......................................296
Refrigerator ...............................................................308
Room air conditioner ..................................................313
Room heater ..............................................................311
Rotor .........................................................................222
Rule, folding ..............................................................20

## S

Saw, compass ............................................................20
Screwdrivers .............................................................12
Service equipment, installation ...................................137
Short circuits .............................................................65
Smoke detection alarms ..............................................253
Soldering ..................................................................44
Soldering iron ...........................................................19
Sump pump ...............................................................314
Surface metal molding, installing .................................93
Surface-mounted
    fire/security systems ............................................256
Switch applications ....................................................103
Switches, four-way .....................................................106
Switches, three-way ...................................................105
Switches, three-way, installation .................................108
System, electrical .......................................................25
Systems, raceway .......................................................51
System, typical ...........................................................26

## T

Tools, hand ................................................................160
Tubing, installing .......................................................266
Tubing system, planning .............................................264

## V

Vent fan ....................................................................315
Voltage tester ............................................................17

## W

Wall switch, installing in
    plastered wall ....................................................96
Wall toaster....................................................316
Waste aerator ................................................312
Water pump ....................................................312
Wire connections ............................................41
Wire splices, types .........................................35
Wires, cleaning ...............................................30
Wires, stripping...............................................30
Wiring, concealed ...........................................73
Wiring systems ...............................................52
Work safety considerations ............................21
Wrench, adjustable.........................................14